Water is Life

Questions to Consider

- What is water, and why is water important?
- What are the special properties of water, and why are they important?
- What is the hydrosphere, and what is its importance to life on Earth?
- How do we use water? How much water is available for human use?
- What are natural resources? What are aquatic resources?
- What is conservation? Why is it important?
- How can we tell if water is polluted or clean? How does water pollution affect aquatic life?
- How does the temperature of water affect the amount of oxygen in it?
- What is water quality? How do humans affect water quality?

CHALLENGE QUESTION

What is the source of your drinking water? Are there any water quality or water quantity issues about your water source? What will affect your water source in the future? Who makes decisions that affect your water source?

Did you know that plants and animals are made up mostly of water (fig. 1.1)? Humans need clean water to keep them healthy. Without drinking water you would die in about one week. Our need for water links us to the past and to other living creatures. Since ancient times, societies have succeeded or failed according to their ability to get clean water and protect it for future use.

Figure 1.1. Your body is about 75% water. Photograph courtesy of Texas Parks and Wildlife Department.

Today's modern cities still depend on water for everything from flushing toilets, to fighting fires, to building the cars we drive. Your community depends on water, too. Using water wisely is critical to our survival. Water sustains life on Earth. Life is impossible without water.

What Is So Special about Water?

Water is made up of molecules that have two hydrogen atoms and one oxygen atom joined together (fig. 1.2), so we also call water H_2O (pronounced "h-two-oh"). Water can take three forms: liquid, solid (ice), and gas (vapor). Water exists in all three forms on Earth.

Figure 1.2. Water molecules are called H_2O (pronounced "h-two-oh"), because two hydrogen atoms (H) attach to a single oxygen atom (O). Illustration by Rudolph Rosen.

(A) The molecule is held together by the attraction of oxygen and hydrogen atoms to each other, because the two atoms have opposite charges. These charges also strongly attract other water molecules like a magnet, playing an important role in the special properties of water.

(B) As water freezes, water molecules align into a regular structure.

(C, D, E) As ice is formed, it expands in size and becomes less dense than liquid water. This characteristic explains why water pipes in a house burst in freezing weather and why ice floats on water.

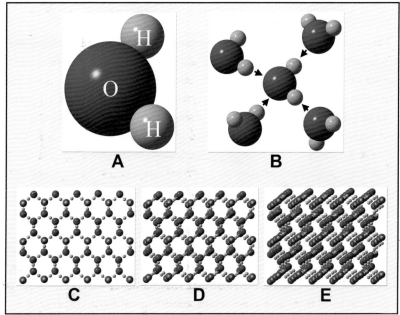

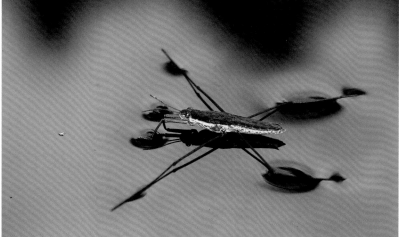

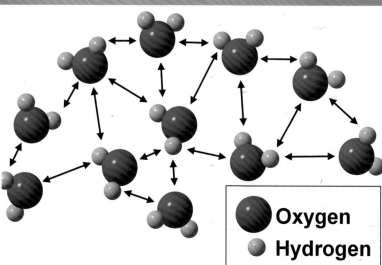

Figure 1.3. Water's strong attraction to itself, called cohesion, creates surface tension that allows insects such as the water strider to walk on it without breaking through. The water molecules at the surface are more strongly attracted to other water molecules on the surface than those below the surface because there are no water molecules directly above the water molecules on the surface. This forms a strong surface "film." It takes more force for an object to break through the surface film than to move through the water below it. Photograph courtesy of Valerie Bugh, larvalbug.com; illustration by Rudolph Rosen.

Water can travel great distances. For example, clouds are made up of tiny ice crystals and water droplets. Water can move up plant stems, keep animals and plants alive, break rocks, and dissolve or erode almost anything. The chemical structure of water gives it these amazing powers. Water molecules attract one another in a way that makes them form drops. Have you ever watched raindrops creep across a window? When the droplets get close to each other, they join to form one larger drop. Water molecules cling to other things, too. This ability to cling allows water to move up plant roots and enables blood to flow through tiny blood vessels. It also enables bugs and other small things to rest on the surface of water without sinking (fig. 1.3).

Water boils at 212 °F (100 °C) and freezes at 32 °F (0 °C). Water can absorb a lot of heat before its temperature begins to rise. A large body of water heats up very slowly, and it cools down just as slowly. This property allows living things to survive in water's fairly constant environment. Water is unusual because its solid form (ice) is not as dense as its liquid form. This explains

Figure 1.4. Erosion from water flowing across land can carve away entire mountains over time. The Big Bend area of the Rio Grande is an example of a landscape in Texas eroded by water. Photograph courtesy of Texas Parks and Wildlife Department.

why ice forms at the top of a lake, floating on the unfrozen water below it and insulating aquatic organisms from extreme temperatures. Without this property, lakes would freeze solid throughout, trapping and killing fish and other aquatic life.

Chemists call water the "universal solvent" because it is very good at dissolving many different things. Seasonal cycles of rainfall and flooding have shaped many landscapes through erosion (fig. 1.4). Because of these properties, water in nature is never completely pure. This also means most of the chemistry of life happens in solutions of water. For example, blood is really just a lot of our cells contained in water! Water also contains dissolved oxygen and other gases from the air, dissolved minerals from the Earth, and **organic matter** that supply essential elements and nutrients needed by aquatic organisms to live and grow underwater. Unfortunately, the gases, minerals, and other things that water dissolves can also pollute it. **Water pollution** occurs when there are too many natural or human-made substances of a harmful nature in the water. These substances can harm or kill the plants and animals that live in or near the polluted water.

How Much Water Is There?

At this moment, the Earth has all the water it has ever had or will ever have. Our planet will get no new water. The good news is that water is rarely ever destroyed. Earth's water is continually being recycled through the hydrologic cycle. From wetlands, oceans, and lakes, water evaporates into the atmosphere as water vapor and then falls back to Earth as rain. Water cannot be made in a factory.

The **hydrosphere** is all the water on Earth. Water covers about 71% of the Earth's surface. That is about 358 quintillion (358,000,000,000,000,000,000)

Figure 1.5. Less than one-half of one percent of Earth's water is available for use by people. Illustration courtesy of Missouri Department of Conservation.

gallons of water! But 97% of the world's water is ocean water, which is **saltwater** and too salty for humans to drink. That leaves only 3% of the water as **freshwater** to supply the whole world. Yet most of that freshwater, about 80%, is frozen in the polar ice caps. Most of the rest of the freshwater is too polluted to use, is trapped in soil, or is just too hard to get (fig. 1.5). That leaves us with just one-half of one percent of the world's freshwater to use. As the world's population grows, the demand for freshwater grows. Luckily, water is also one of the most recyclable substances on Earth. In fact, all the water on the planet has been recycled countless times—we drink the same water the dinosaurs drank! So when we use water, we do not destroy it or make it disappear. We may move it. We may make it unusable for drinking purposes, sometimes for a short period or longer if water becomes polluted with harmful **contaminants**. The more polluted it becomes, the more difficult it may become to clean it. But eventually water is recycled.

A person can live on a gallon of water a day for drinking, cooking, and washing. But most people in Texas use far more than this. On average, every American uses about 90 gallons of water each day. Worldwide the need for water has tripled over the past 50 years. To get this water, people have drained rivers dry, turned grand valleys into huge tubs of water, and pumped so much water out of the ground that the Earth's surface has sunk beneath our feet. In the United States, we often take for granted the clean water that flows out of our faucets. We can assume our water is safe. We do not think much about where it comes from, how the water is cleaned and distributed, how much we use each day, or what happens to it when we flush our toilets.

Waste Not, Want Not

Compared to many countries, the United States is water rich. We have 39 million surface acres of lakes and reservoirs. The Great Lakes cover about 6.7 million acres and contain about one-fifth of the world's freshwater supply. Water covers about 4% of the United States. This abundance has allowed the United States to grow surplus crops and build profitable industries. Everyone uses water, but it may surprise you to learn that only about one-tenth of the water used goes to our homes and cities.

Figure 1.6. Irrigation, or watering of crops, to grow food is the largest use of water in the United States. Photograph by Rudolph Rosen.

Agriculture is the biggest user of water (fig. 1.6). For example, it takes about 24 gallons of water to grow one pound of potatoes. To produce one pound of beef takes about 2,600 gallons of water. Growing a day's food for one adult takes about 1,700 gallons of water. The second biggest use of water is for industry. Producing electrical power takes more water than any other industrial use, but almost all the water used is returned to or never removed from its source. Moving water where we want it to go actually uses power, too. Approximately 4% of the nation's electricity is used just for moving and treating drinking water and wastewater. Did you know that it takes 3,000 to 6,000 gallons of water each year to power one 60-watt light bulb for 12 hours each day for a year?

Water in Texas

Water is the most important **natural resource** in Texas. We rely on water found on the surface and underground. Texas has more than 1.26 million acres of freshwater in lakes, ponds, and reservoirs. There are an additional 2.1 million acres of water in the bays and estuaries along the Texas coast. There is still more water in Texas' 191,000 miles of streams. Texas is second only to Minnesota in the surface area of lakes and reservoirs and second only to Alaska in total volume of freshwater. Over three-quarters of the state sits over an enormous underground aquifer. Water in Texas bubbles up from about 3,000 springs, including some of the largest in America. Texas is also unique in not having just one or two large rivers, as most states do. More than 3,700 streams feed into 15 large river systems in Texas. The rivers provide water for our personal use, agriculture, communities, recreation, transportation, industry, and fish and wildlife (fig. 1.7).

Figure 1.7. Texas' rivers and lakes provide fun recreational opportunities like fishing, kayaking, and swimming. Shown here is stand-up paddleboarding in Lady Bird Lake, Austin. Photograph courtesy of Texas Parks and Wildlife Department.

If Texas is so rich in water, why should we be concerned about its **conservation**? The answer is that some parts of the state are water rich while other parts have very little water and can be affected by **drought**. For example, South and West Texas are **arid** parts of the state. They can be very dry, even when other parts of the state may be very wet. El Paso, on the extreme western edge of the state, is part of the Chihuahuan Desert and usually receives less than 10 inches of rain per year. Port Arthur, on the eastern edge, can receive more than 60 inches of rain per year.

Most of Texas has modest rainfall compared to that in the rest of the country. Several years of low rainfall can leave entire parts of Texas without water. In addition, people can cause harm to the water in lakes and rivers by pollution and careless use. This in turn harms **aquatic resources**, such as fish, and can make water undrinkable and useless for recreation. To protect our aquatic resources, we must use them wisely. Conservation means careful use. That is what this book is all about.

Pollution Kills

Pure water is clear and transparent to light. As long as water is relatively clear, plants can live under the water, using energy from sunlight passing through the water to make food through photosynthesis. Sometimes water can be discolored from small amounts of natural organic matter in the water, but the water may not be polluted. It can be transparent enough to allow light to penetrate. But this same organic matter that may discolor water at low levels may become harmful to aquatic life at high levels. Water that looks "dirty" may be polluted or just dirty. **Turbid** water contains suspended material, such as dirt, **silt**, or mud particles. Few plants grow in muddy water because the silt absorbs light. And to make seeing pollution in water even more complicated, not all clear water is clean.

Water may look clean but still be polluted. A body of water may have invisible toxic (poisonous) chemicals in it. Most toxic pollution comes from human-made (manufactured) chemicals that enter water, such as herbicides, pesticides, and industrial compounds. Another kind of pollution we cannot see is too much heat. When water is exposed to air, the water and air mix and water picks up oxygen from the air. This is called **aeration**. You may have seen aerators on aquariums that force air bubbles into water. In nature, flowing water and waves at the water's surface aerate water. This provides dissolved oxygen to plants and animals that live in water. Hot water holds less dissolved oxygen than cool water. The plants and animals we see living in water need oxygen, which is dissolved as a gas into the water along with other gases, such as carbon dioxide. If the water level in a lake or stream gets too low and summer heat warms the water too much, fish and other aquatic life do not get the oxygen they need and die.

Pollution can also occur when too much organic matter, such as manure or human sewage, gets in the water and decays. Decaying organic matter uses

Figure 1.8. Fish, birds, amphibians, reptiles, and all other wildlife depend on water for life. In particular, aquatic resources such as fish and pond turtles require abundant clean water. People benefit when fish and wildlife populations and the habitats in which they live are healthy. Photograph courtesy of Texas Parks and Wildlife Department.

up a lot of the dissolved oxygen in water. Organic pollution can also happen when **inorganic** pollutants such as **nitrates** and phosphates build up in the water. People use nitrates and phosphates as fertilizers on crops, gardens, and lawns because these materials help plants grow. High levels of fertilizing chemicals in the water feed the growth of plants and algae. Too many plants growing on the water's surface can block light from reaching deeper water. Then as plants and algae die and **decompose**, dissolved oxygen in the water is quickly used up. This process of rapid plant growth followed by rotting and oxygen loss can result in death of fish and other animals.

Quality Water Means Quality Life

The amounts and types of pollution in water affect **water quality**, which is water's fitness for a particular use. Water in a pond or river might not be pure enough to drink without special treatment, but it may be just fine for agriculture or industry, or even for swimming or fishing. Water professionals in our communities take thousands of water samples and perform tests each year to determine which chemicals and disinfectants must be used to treat raw water to make it suitable for human use. Municipal drinking water is highly regulated in the United States and safe to drink.

Many things affect water quality. Physical properties such as suspended silt and temperature make a big difference. Chemicals or salts in water also change quality. Water quality tests are used to measure properties of water, such as acidity and salt level. Tests that measure water's ability to conduct electricity can show how much salt and other substances are in the water. Tests can also measure the amount of dissolved oxygen and detect the presence of chemicals such as fertilizers. The presence or absence of plants and animals in a body of water tells a lot about water quality (fig. 1.8). A wide variety of healthy aquatic organisms, including plants, insects, and fish, in the water means that the water quality is good.

AQUATIC SCIENCE CAREER

Water Quality Regulator

The federal Clean Water Act was passed to protect America's waters from pollution and misuse. States also have laws to protect water. Water quality regulators enforce these clean water laws. They work for agencies such as the US Environmental Protection Agency, Texas Commission on Environmental Quality, and Texas Parks and Wildlife Department. Workers in these agencies monitor water pollution and may collect data that result in finding polluters. They test water quality and collect water samples for chemical and biological analysis. These jobs require a college degree in science.

Photograph courtesy of Texas Parks and Wildlife Department.

chapter 2

∎ ∎ ∎ ∎ ∎ ∎ ∎

The Ultimate Recyclable: Water

Questions to Consider

- What is the hydrologic cycle? Where does it start, and where does it end? Where does water spend most of its time?
- What is weather? What is climate? How do they affect the quality and quantity of our water?
- What kinds of climates are found in Texas?
- What is surface water? What is groundwater?
- Where does water go when it runs off a street?
- Where does our water come from? How does it get to our faucets?
- What happens to water after it goes down the drain?

CHALLENGE QUESTION

If water is so recyclable, how can we use this property of water to create a sustainable water future in Texas?

For 3.5 billion years, the Earth's water has been moving from streams to lakes to oceans, flowing underground, sitting high up on mountain glaciers, freezing and melting on the edges of the polar ice caps, and forming clouds in the **atmosphere**. The drop of water that falls on your head in a rainstorm could have traveled from the Pacific Ocean, from a mountaintop wetland in Colorado, or from a melting glacier in Greenland. This never-ending water recycling is called the **hydrologic cycle** (sometimes called the water cycle), and it is the driving force behind our weather (fig. 2.1).

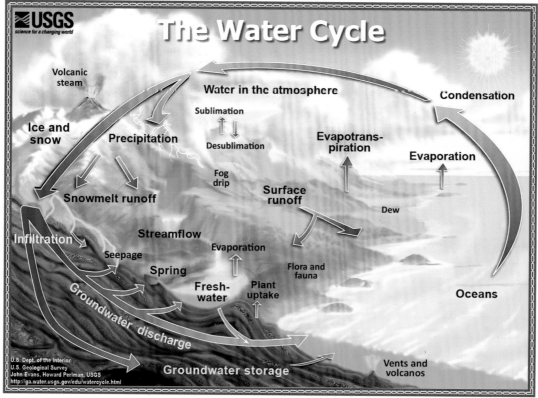

Figure 2.1. Water is constantly recycled, moving from ocean, to mountaintop, into the ground, to the atmosphere, to the ocean, and back again. The water is cleansed of pollutants as it changes from liquid to vapor. Illustration courtesy of US Geological Survey.

Solar-Powered Water Pump

The hydrologic cycle works like a huge water pump powered by solar energy and gravity. It is a global system, and every molecule of water on Earth travels through it. Because it is a cycle, it has no beginning or end. The sun warms water on the Earth's surface and changes it into invisible water vapor. This process is called **evaporation**. Every time water evaporates, it leaves behind whatever salts, pollutants, or other impurities were dissolved in it and becomes pure again. But as soon as the purified water is exposed to air or the Earth's surface, it begins to pick up pollutants again. This explains why even rain can be polluted with contaminants found in Earth's atmosphere. Living things, which make up the **biosphere**, also return water to the atmosphere. Every time we exhale, we release water vapor. Through photosynthesis plants release water vapor into the air in the process of **transpiration**. A one-acre cornfield (about the size of a football field) can give off as much as 4,000 gallons of water every day through the corn plants' leaves, or as much water as you will use to take about 600 showers.

Rising air takes the water vapor up into the atmosphere, where the vapor cools. Cooling water vapor condenses into tiny suspended water droplets as fog, mist, or ice crystals, which we often see as clouds. In fact, a cloud is really just a huge group of tiny water drops held up by rising air (fig. 2.2). Raindrops and snowflakes condense around microscopic dust particles also suspended

in the atmosphere. Water can pick up other contaminants from the air, too, such as smog (forming **acid rain**) or mercury vapor from trash incinerators and coal-burning power plants. Water returns to Earth as **precipitation**, either liquid (rain) or solid (snow, sleet, or hail). This happens when the rising air can no longer hold up all the droplets of water. Water vapor can also condense on ground-level surfaces as dew or frost. About 85% of the world's precipitation falls into the oceans, and the rest falls on land.

Figure 2.2. Clouds form out of condensed water vapor. Tiny water droplets and ice crystals from clouds fall to Earth and fill our lakes and rivers with clean water. Photograph by Rudolph Rosen.

Talk about the Weather

Weather is the condition of the atmosphere at any point in time, such as if it is hot or cold, raining or dry, windy or calm, or clear or cloudy. The amount of water that falls in a local area changes with each season. Weather also increases or decreases the amount of available water. Seasonal weather patterns move water around the world and from the atmosphere back to the Earth's surface. **Climate** is the average weather conditions over time. How much water there will be in a certain region in a given part of the hydrologic cycle depends on factors such as these:

- amount of rainfall
- effect of temperature on evaporation
- amount of water that plants use during the growing season

Weather in Texas

Texas is so large that it is affected by air drawn in from the Pacific Ocean as well as the humid tropical air flowing in from the Gulf of Mexico. The state's

ULTIMATE RECYCLABLE

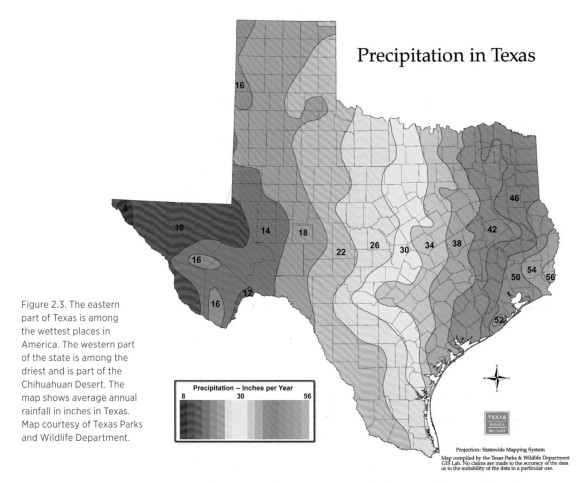

Figure 2.3. The eastern part of Texas is among the wettest places in America. The western part of the state is among the driest and is part of the Chihuahuan Desert. The map shows average annual rainfall in inches in Texas. Map courtesy of Texas Parks and Wildlife Department.

vast size also means climate varies greatly from north to south and east to west (fig. 2.3). Rainfall amounts increase from the western to the eastern part of the state. The wet season does not happen at the same time everywhere. North, Central, and East Texas receive the most rain in late spring, while the Panhandle and West Texas receive the most rain during the hottest months of the year. Coastal areas of Texas receive the most rain in late summer and early fall.

Most of Texas has a modified marine climate, which means it is influenced by the onshore flow of tropical air from the Gulf of Mexico. From October to June this tropical air, which is full of moisture, collides with colder dry air from the north, resulting in thunderstorms over Central Texas. Along the coast, hurricanes can bring huge rainstorms with large amounts of rain falling over short periods. This may cause flooding, water surges in bays and estuaries, and high winds. Peak time for hurricanes is late summer and early fall. Texas is known for its extreme flooding. Flooding is natural but can cause major damage to buildings, homes, and roads, as well as damage to aquatic resources and habitat for wildlife.

Although droughts do not happen as fast as floods, they may also cause severe economic damage and harm aquatic resources and wildlife habitat (fig. 2.4).

Damage to property and habitat can be reduced by preserving wetlands and allowing natural sediment transport and adequate freshwater inflows into our estuaries and bays.

Water also plays a role in temperature, which varies throughout the state: the coldest temperatures are in the north, and the warmest are in the southwest (an area known as the Trans-Pecos). In dry areas the difference between the highest and lowest temperatures is larger than in more humid areas. Water in the atmosphere, called **humidity**, buffers large changes in daily temperature. For this reason, even if Houston and El Paso have the same daytime temperature, Houston will typically be warmer than El Paso at night because the higher humidity in Houston will keep the air warm.

Weather forecasters sometimes refer to the effects of El Niño and La Niña. These are changes in ocean-atmosphere conditions that can happen every few years and affect weather around the globe. El Niño occurs when the area of warmer waters in the tropical Pacific Ocean expands, causing warmer, wetter, and more severe weather conditions in Texas. La Niña occurs when the area of warmer waters shrinks in the Pacific Ocean, causing cooler, drier conditions in Texas.

The National Oceanic and Atmospheric Administration (NOAA) notes that El Niño is occurring more frequently. Is there a connection between this frequency and climate change? Will we have more severe weather as a result? According to NOAA, these are the important research questions facing

Figure 2.4. Texas is at the same latitude as the Sahara, the world's hottest desert. As they do in the Sahara, large high-pressure cells in the atmosphere may sit over Texas for weeks or months, blocking storms and moisture from reaching the state from any direction and thus increasing drought conditions. Photograph courtesy of Texas Parks and Wildlife Department.

scientists today. Many scientists believe we are now encountering a period of rapid climate change. Overall, Texas is known for extremes in weather, from droughts to floods, from freezing temperatures to burning heat. All are present and relatively common in our state.

Surface Water Runs Off

On land, plants catch much of the rainfall before it reaches the Earth's surface. In a wooded area, for example, rain slowly drips off leaves of bushes and trees and trickles down branches. Roots and the leaf-covered forest floor act like a sponge, soaking up water and slowly releasing water into the ground or waterways. About 66% of the precipitation that falls on the continental United States each year returns to the atmosphere right away. Half of the rest runs off the surface of the land. This water is called **surface water**. It collects in streams and flows to the ocean. People and some wild-

Figure 2.5. Beaver and some other wildlife species have learned ways to save water for their needs, much as people build dams and reservoirs to store water for drinking, agriculture, power production, and recreation. Photographs courtesy of Texas Parks and Wildlife Department.

life species, such as beaver, may build **dams** to slow and hold water for use later (fig. 2.5).

When precipitation falls as snow, it can build up as snowpack, ice caps, and glaciers. Ice caps and glaciers can store frozen water for thousands of years. Snow in warmer places melts when spring arrives. The melted water, or snowmelt, that does not soak into the ground flows overland.

If rain is hitting the ground faster than it can soak in, it becomes **runoff**. The slope of the ground also affects runoff. On steep slopes, water moves quickly and very little of it soaks into the ground. Hard surfaces reduce the amount of water that soaks into the soil even further. In urban areas, there is less **porous** ground for the rain to soak into. Paved roads, rooftops, and parking lots block water from soaking in, so all of it becomes runoff. Heavy rains run off streets, sidewalks, and other paved surfaces up to 10 times faster than on unpaved land. Some people use rainwater harvest barrels or tanks to collect water as it falls from building roofs. The water can then be used to water plants during dry periods when rain is scarce. This is a good way to conserve water.

The faster water flows, the more power it has to wash away soil or to cause flooding (fig. 2.6). Storm water that runs off paved roads, rooftops, and parking lots flows into ditches and storm drains. This water then drains directly into streams, lakes, and wetlands without any filtration or treatment. Any excess fertilizer, pesticides, mud, motor oil and antifreeze, trash, lawn clippings, and pet waste on the pavement or roads wash into waterways during heavy rains. This type of runoff is called **nonpoint source pollution** and is the leading cause of pollution of our aquatic resources in the United States.

Figure 2.6. Flooding occurs when there is more runoff than can be absorbed into the ground or fit into streams, rivers, lakes, and reservoirs. The water has nowhere to go so spills out into adjacent areas. Photograph courtesy of Corpus Christi Caller-Times.

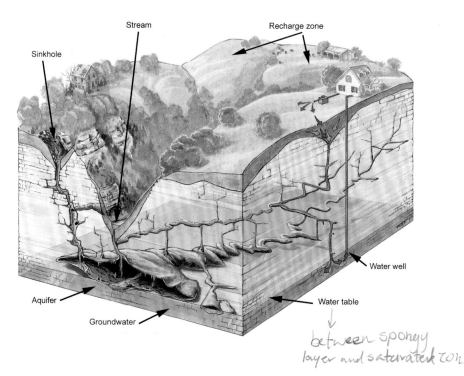

Figure 2.7. Groundwater is recycled slowly and can be depleted if more is pumped from the ground by people than soaks into the ground each year. Many people depend on groundwater for drinking water and irrigating lawns and crops. Illustration courtesy of Missouri Department of Conservation.

Let It Soak In—Groundwater

Only about 3% of rain soaks into the ground. When water soaks into the ground, it fills the empty spaces between soil particles. The water may remain as soil moisture, evaporate back into the atmosphere, be taken up into the roots of plants, or trickle slowly through the soil. The solid part of the Earth is called the **geosphere**.

Below the Earth's surface, layers of spongy soil, sand, and rock act as filters to help clean the water. If the water is badly polluted or contains certain contaminants, the soil cannot remove all of the pollutants. In some cases, water moving through the geosphere can even pick up pollutants already present in the soil. Eventually the water reaches a layer where all the spaces in the soil or rock are already filled up. This area is called the **saturated zone**, and the water it holds—over half of the freshwater on Earth—is called **groundwater** (fig. 2.7). The boundary between the spongy layer and the saturated zone is known as the **water table**. The water table rises or falls as the amount of groundwater in the saturated zone increases or decreases.

Areas of underground rock that hold water in pores or crevices are called aquifers. To use the water in aquifers, people dig wells to bring it to the surface. Unfortunately, digging a well and pumping out too much groundwater can lower the water table and can cause lakes, streams, and wetlands to dry up. Sometimes the Earth's surface can sink or even collapse when we pump out too much groundwater. In most places, groundwater moves so slowly that aquifers can take thousands of years to fill back up. In the Texas Hill Country, however, the rate of flow is much faster and in some places can be measured in miles per hour. Places where water soaks into aquifers are recharge areas. Some streams lose water to the soil or rock around them, helping recharge aquifers. Streams also receive groundwater with the most visible examples called springs. People have long inhabited lands near springs because of the ready availability of a constant supply of freshwater (fig. 2.8).

Clean Water for Us

Texans depend on both surface water and groundwater sources for drinking and other uses. In general, people in the western portion of the state use groundwater. In wetter parts of the state large reservoirs hold surface water that can be added to groundwater sources for people to use. Some well water is safe to drink right out of the ground. In other cases, it must be treated first. For community water supplies, water from wells or lakes is piped to drinking water treatment plants. At the plants, workers may filter the water or let it settle to remove suspended material. They also may add chemicals to the water to kill bacteria and other organisms. Pipes buried in the ground carry the water to homes and businesses.

Clean Water for the Environment

A different set of underground pipes carries used water (sewage) to **wastewater treatment facilities**. Sewage from homes and businesses that are not located in towns or cities usually goes to underground septic systems that store wastewater until it can soak into the ground. The solid material is disinfected and often recycled into fertilizer. Wastewater treatment systems

Figure 2.8. The San Marcos Springs, on the campus of Texas State University in San Marcos, is the second-largest spring in America west of the Mississippi River. The springs are home to eight endangered aquatic species and are thought to be the longest continually inhabited site by humans in North America. Visitors can see the springs bubbling from an underground aquifer and many species of aquatic life by taking glass-bottom boat rides. Photograph by Rudolph Rosen.

ULTIMATE RECYCLABLE

use bacteria to break down much of the organic waste in the water and make the wastewater safer to return to the environment. Water from sewage treatment facilities is piped back into surface waters such as rivers, lakes, and wetlands after it is treated. Wastewater treatment is important to the reduction of water-borne diseases that can sicken humans and even cause death.

AQUATIC SCIENCE CAREER

Wastewater and Drinking Water Treatment Plant Worker

Wastewater treatment plant workers process wastewater so it is safe to return to the environment. They run equipment that removes or destroys contaminants, bacteria, or other harmful pollutants in the water. They also control pumps and other machinery that move wastewater through the treatment processes. They may investigate sources of pollution to protect people and the environment.

Drinking water treatment plant workers make water safe to drink. They read and adjust meters and gauges to make sure equipment is working properly. Water from wells or lakes is piped to the treatment plant, where workers may add chemicals to the water. They take samples of the water and analyze them. These jobs require special training, certification, and often a bachelor's degree.

chapter 3

■ ■ ■ ■ ■ ■ ■

What Is Your Watershed Address?

Questions to Consider

- What is a watershed? Which watershed do you live in?
- When it rains at your home, to what creek and river does the water travel?
- How does the watershed affect the water body into which it drains? How do human activities affect the quality and quantity of water in a watershed?
- What is point source pollution? What is nonpoint source pollution? What are examples of each?
- What is erosion? What causes it?
- What is sediment? Where does it come from?
- How does human activity affect erosion and sedimentation? What is the impact of erosion and sedimentation on aquatic resources?
- What are the natural regions of Texas? Is the water in your region drinkable, swimmable, and fishable?

CHALLENGE QUESTIONS

How does your location within a watershed affect the quality of water where you live? How can you influence the quality of water for others in your watershed?

It is usually difficult to see a watershed unless you are standing on top of a hill or looking down from an airplane. Then you can see all the hills and valleys that drain water into a stream, the streams flowing into rivers, and the places where rivers flow into lakes, bays, and estuaries (fig. 3.1). A **watershed** is all the land from which water drains into a specific body of water. Sections

Figure 3.1. Watersheds are all the land from which water drains into a particular body of water, such as a stream, lake, or ocean. Illustration courtesy of Missouri Department of Conservation.

of connected hills and valleys form each watershed, and small watersheds make up a larger watershed. Everyone lives in a watershed, which is the land upstream from you, and all land on Earth is part of some watershed. As you move downstream, more streams and rivers flow together. Then all the new land drained by those new streams adds to your watershed. So your watershed depends on where you are along a stream, river, lake, estuary, or ocean.

Flowing to a River

If you stand on top of a mountain that divides two watersheds, you can pour a glass of water from one hand into one watershed and a glass of water from the other hand into a different watershed. Sooner or later, the water from the two glasses will end up in two different streams, which combine with other streams to form rivers. The area drained to form the river is a watershed but may also be called a **river basin**. The river basin is the land drained by a river and its **tributary streams**. All rivers eventually flow into the ocean.

Your Watershed Address

Most states have only 1 or 2 river basins, but Texas has 15 major river basins and 7 major bays. Texas also has more than 3,700 named streams. All the streams and rivers combined flow over 191,000 miles through Texas. The Red River forms the border between Texas and neighboring states to the north, and the Sabine River forms our border with Louisiana to the east. The Rio Grande, the second-longest river in the United States, forms our border with Mexico in the south and west. All rivers in Texas eventually flow to the Gulf of Mexico. As several of our major rivers get closer to the Gulf of Mexico, they combine into **coastal basins** and form major bays and estuaries. Your **water-**

shed address is the watershed, sub-watershed, sub-sub-watershed, and so on, in which you live. It tells which lake, stream, or wetland collects the water that falls on your home (fig. 3.2).

At the top of the watershed is the land known as the **headwaters**. This is the high ground where precipitation first collects, or it can be a spring from which a stream originates. From the headwaters, water flows downhill and eventually forms a permanent channel that is called a stream. Small streams combine to form larger streams. The mouth of a stream is the place where it empties into a larger body of water. A permanent or **perennial stream** is one that has flowing water in it all year long. In the Texas Hill Country and other dry parts of the state, many streams flow for only part of the year, although water may be flowing just below ground, beneath the **dry streambed** (the bottom of a stream). Such streams are called **intermittent** or **ephemeral**.

Figure 3.2. Texas major river basins, coastal basins, and major bays. Map courtesy of Texas Parks and Wildlife Department.

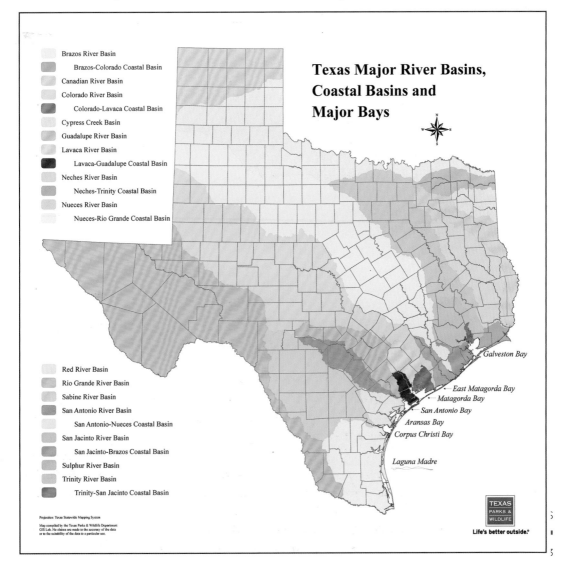

What You Do to the Land, You Do to the Water

Everything that happens on the land in a watershed affects the water body into which it drains. A stream, pond, or wetland can be only as healthy as its watershed. How we use the land affects the health of our aquatic resources and in turn affects us and what we do in the watershed, for example, fishing. In a healthy watershed, water is filtered and stored, but as water runs downhill, it can pick up whatever is on the ground, beginning a process of altering water quality. When it flows through cities or across fields and pastures, water picks up dirt, pollutants, and heat (fig. 3.3). These contaminants flow into a stream, wetland, or lake, affecting the water you use to drink, swim, or fish. When you flush your toilet, do the laundry, fertilize your lawn, or dump used oil on the ground, you are affecting water quality in your watershed and all the way into the Gulf of Mexico.

Water pollution is described in two ways. One way is **point source pollution**, which comes from a single source that people can identify (or point to), such as a pipe connected to a specific place. The other way is **nonpoint source pollution**, which comes from a combination of many sources rather than a single outlet. Examples of nonpoint source pollution include runoff from fields and construction areas; fertilizers used on lawns and golf courses; fuel, oil, and antifreeze from roads; and animal waste and bacteria from feedlots.

Finding and preventing water pollution in our state are vital to every Texan's quality of life. What do you think is the biggest pollutant in Texas waters? Sewage? Industrial chemicals? Pesticides? Fertilizers? Trash? Believe it or not, the biggest pollutant is plain old dirt from excess **sedimentation**, even though chemical pollution and too much bacteria in the water are what people usually hear about the most (fig. 3.4).

The proper name for dirt is soil, and when it gets into a water body, it becomes **sediment**. Sediment is any bit of rock or soil, such as mud, clay, silt, sand, or gravel—even boulders. Excess sediment blocks out light, killing

Figure 3.3. Many streams and rivers run through Texas cities. Here the San Antonio River runs alongside San Antonio's famous River Walk. Photograph courtesy of Texas Parks and Wildlife Department.

Figure 3.4. Sedimentation is the greatest source of pollution of rivers in Texas. Photograph courtesy of Texas Parks and Wildlife Department.

aquatic plants or preventing their growth. Sediment covers up the nooks and crannies where aquatic organisms live. It smothers fish by clogging their gills and by reducing the amount of dissolved oxygen in the water. How can we reduce the amount of sediment in our rivers?

Raindrops Can Move Mountains

Raindrops fall at a speed of about 30 feet per second, or 20 miles per hour. When a raindrop strikes bare soil, it creates mud that is splashed as much as 2 feet high and 5 feet away. This is an example of **erosion**, the wearing away and movement of solid material such as soil, mud, and rock. It is a very powerful natural process caused by the forces of wind, water, ice, gravity, and living things (see fig. 1.4).

Erosion and sedimentation in water are natural processes. However, too much of either can cause problems. Erosion can remove fertile soils from farmland and reduce quality of water. Sediment that erodes from one place is carried away and settles out downstream. This can clog streams with gravel and fill reservoirs with sediment. Erosion can be a big problem in areas where too many trees are cut down, where poor farming practices leave the ground bare, or where construction exposes bare soils to rain and wind. Planting trees and grasses can slow erosion. Leaves and stems slow the rain as it falls to the ground, and plant roots hold soil and rock in place. Many Texas farmers and ranchers have worked to improve farming and grazing practices and reduce the amount of soil and other sediment in Texas streams.

We can often see the close tie between land and water. Texas has 11 major **natural physiographic regions** (fig. 3.5). These regions have different types of bedrock, soil, elevation, weather, and plants. These differences and the different ways the land is used in each affect the overall water quality and

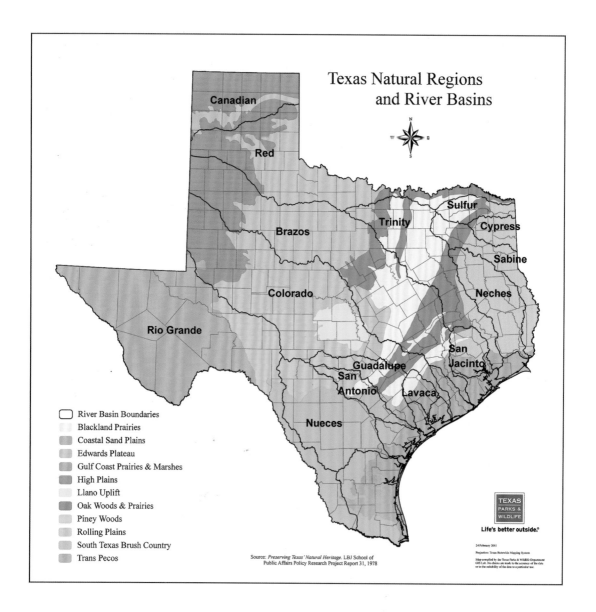

Figure 3.5. Texas natural regions and river basins. Map courtesy of Texas Parks and Wildlife Department.

quantity in the watersheds. Each region has different kinds of habitat for wildlife and opportunities for people (fig. 3.6). Every stream, lake, or wetland is a reflection of its watershed. The goal of the Clean Water Act is water that is "drinkable, swimmable and fishable." Natural resource agencies, communities, and individuals work together for good water quality and quantity. Knowing our watershed and its relationship to surrounding watersheds can help us conserve our aquatic resources.

CHAPTER 3

AQUATIC SCIENCE CAREER

Hydrologist

Hydrologists study the movement, distribution, and quality of water. They test, measure, and collect water data, such as river flow rate, tidal fluctuations, dissolved oxygen, sediment load, acidity, salinity, and groundwater levels. These data help us learn about the oceans, surface water on land, and groundwater in our aquifers. Hydrologists write reports; prepare water maps, tables, and graphs of study results; and perform data analyses. These are published in documents or scientific journals and can be used to support water projects or investigations. Hydrologists have at least a bachelor's degree; many have a master's or doctorate degree.

Photograph courtesy of Harte Research Institute for Gulf of Mexico Studies.

Figure 3.6. The natural physiographic regions in Texas can provide very different boating and fishing experiences. Canoeing in Caddo Lake in East Texas (*left*), paddling in Aransas Bay near Rockport (*center*), and kayaking at Parrie Haynes Ranch on the Lampasas River near Killeen (*right*). Photographs courtesy of Texas Parks and Wildlife Department.

chapter 4

■ ■ ■ ■ ■ ■ ■

Living in Water

Questions to Consider

- What is a species?
- What is an adaptation?
- What adaptations do fish and other aquatic animals possess to survive in an aquatic habitat? How do specific adaptations provide survival advantages to a particular species?
- What are some adaptations of different fish species in Texas?
- How do fish swim?
- How do fish see, smell, hear, taste, and feel? Do fish have other senses that we do not have?

CHALLENGE QUESTION

How might aquatic species in Texas differ from species in other states, for example, in Oregon, Michigan, or New York?

You walk beside a pond. A mallard duck flies overhead. A bluegill swims in the water. Humans, mallards, and bluegills are distinct species. A **species** is a group of individuals sharing some common characteristics or qualities and whose offspring also share those characteristics or qualities. All species are specially suited for the lives they lead. Humans are walkers with legs, allowing for effective movement on land. Mallards are fliers with wings that allow them to move through the air. Bluegills are swimmers with fins that enable them to swim in water. An **adaptation** is a **behavioral trait**, **structural trait**, or physiological trait that increases a species' chance of survival in a specific environment. Every living thing has adapted to fit with where it lives in order to survive. **Aquatic organisms** live in water and have adaptations to do so.

Fish Guts

Fish are **ectotherms**, which means that their body temperature changes as the surrounding water temperature changes. The body temperature of most fish remains very close to the temperature of their water habitat. Because temperature has such a great effect on fish, different species have different water temperature preferences. Some species prefer warmer water than others. Most Texas species are warm-water fish, because most of Texas' waters are warm. Fish that live farther south toward the equator prefer even warmer water.

Fish have many of the same internal organs as humans and other mammals. They have a heart to pump blood, intestines and stomach to digest food, kidneys, a liver, a gall bladder, and a spleen (fig. 4.1). **Osmoregulation** is a physiological adaptation found in fish that enables some fish species to live in freshwater and others to live in saltwater. In this process fish regulate their intake of saltwater or freshwater to keep their fluids, such as blood, from becoming too salty or too dilute. Fish living in saltwater have internal fluids lower in salts than the water in which they live. These fish must drink large amounts of saltwater and excrete small amounts of fluids while they actively secrete high amounts of salts through their gills. Fish in freshwater have higher salt concentrations in their body fluids than in the surrounding water, causing them to excrete large volumes of fluids with low salt content and take up salts through their gills. Some species are able to adapt to a wide range of salinity, or salt content. Fish that can live in freshwater, brackish water, and saltwater are called **euryhaline**. Examples of common euryhaline aquatic species in Texas are red drum and blue crab.

Fish have some organs that humans do not have that allow them to live in water. Instead of lungs, fish have **gills**. Gills contain capillaries (fine blood

Figure 4.1. Fish have many of the same organs as humans but also have gills, a swim bladder, lateral lines, and fins that allow them to live in water. Illustration courtesy of Missouri Department of Conservation.

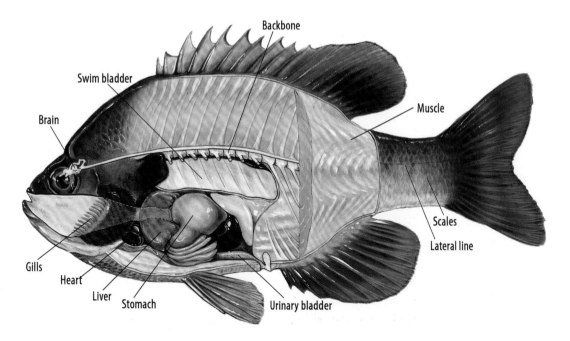

vessels) that take up dissolved oxygen and release carbon dioxide. This occurs as water enters through the fish's mouth, passes over the gills, and exits from the fish. In clean, moving water, a fish can absorb up to 85% of the dissolved oxygen available in well-oxygenated water. Fish and amphibians are the only **vertebrates** (animals with a backbone) that are able to live their entire lives completely submerged in water.

How Fish Swim

Only 30% to 40% of human body weight is muscle, but up to 80% of a fish's body is made of muscle. Fish muscles are packed along its sides, where a fish gets most of its swimming power. When a largemouth bass wants to move forward, it begins a side-to-side wiggle that starts at its head and moves backward along its body. The wiggle pushes water behind the fish, which propels it forward. Fish also use their many **fins** to move about in the water. They have two sets of paired fins (pectoral and pelvic) along the side. They also possess a single caudal and anal fin. Some fish species have a single dorsal fin, while others have two. Certain fish (such as freshwater trout and catfish) also have an adipose fin located on their back, behind the dorsal fin (fig. 4.2).

The dorsal fin, located along the back of a fish, helps keep the fish upright and stable. Some species of fish, such as sunfish, have sharp spines in their fins. These help discourage other fish from eating them. Located underneath the fish near the anus, the anal fin also helps with stabilization. The caudal fin or tail fin can be rounded, forked, or crescent shaped. This helps with speed and movement. Most fish use their pectoral and pelvic fins, which are located along their sides, to steer or maneuver. These fins can be moved independently, giving the fish the ability to move quickly in any direction. Fins can be used as brakes or rudders to help the fish stop, turn, go up or down, or even go backward.

The fish's body shape also affects how it is able to maneuver through water. Certain body shapes may help a fish survive by allowing it to move in and out of tight places to catch food or to escape from predators. Fish with a flat body shape mostly live and feed on the bottom. Torpedo-shaped fish are built for speed. Fish with a tall and thin shape can easily slip in and out of tight places.

Sink or Swim

Many freshwater and saltwater fish have **swim bladders**. Most of the time, the fish uses its swim bladder to keep from sinking. Being able to float or rise in water is referred to as buoyancy. The swim bladder works a little like a hot-air balloon. The more gas (oxygen) it contains, the higher a fish will suspend or float in the water. Some species of fish can also use their swim bladder to make sounds to communicate during courtship, to defend territory, or to sound an alarm when disturbed.

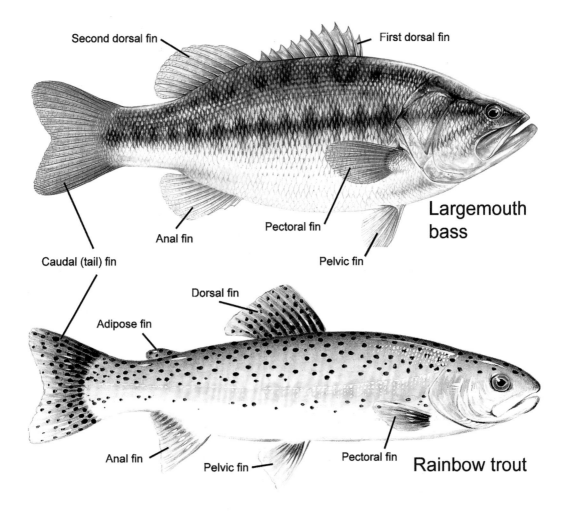

Figure 4.2. Fish fins. Images courtesy of Texas Parks and Wildlife Department; illustration by Rudolph Rosen.

Fish Skin

Many species of fish are covered with **scales** that protect them like roof shingles protect a house. There are four main types of scales with many variations to these. Each type can vary greatly in size. Some fish have scales modified into special purposes, such as the tail spines of stingrays. Fish do not grow more scales as they get older; the scales just get larger. Fish biologists can estimate the age of some fish by counting rings on a scale, similar to the way foresters can tell a tree's age by counting its growth rings. Fish skin is often coated with slime, which helps reduce friction as fish swim through the water. The slime also helps protect them from disease.

Many fish species have structural adaptations such as color or patterns to allow them to blend in with their surroundings. For example, countershading, such as being dark-colored on top and light-colored underneath, helps them blend in with the dark bottom when seen from above, and with the bright surface when seen from below. And you must look very closely to see a darter sitting on a gravel streambed because of its ability to blend into its multicolored surroundings.

Fish Sense

Fish have senses like we do to see, hear, smell, taste, and feel. The senses of some fish are better developed than those of others. Some fish use their sense of smell or taste to find food. Others feed primarily by sight. The placement and shape of eyes of some species allow them to see almost all the way around their bodies. Although fish are nearsighted (cannot see far away), it is difficult to sneak up on a fish because they see in all directions. They can also see colors. Fish that feed at night or live on the bottom in very deep or turbid water rely heavily on their senses of smell and taste instead of sight.

Fish can also hear, and they have a special row of sense organs called the **lateral line** that give them the ability to hear low-frequency sounds. A fish's ears are located beneath the skin on either side of the head. The lateral line is used to detect movement and vibration in the water. Special hairlike cells along the lateral line, either freestanding or located inside very small fluid canals, are very sensitive to vibrations in the water. This allows a fish to sense the watery world around it. The lateral line can be seen on many fish as a faint line running along the side of the body. Lateral lines serve an important role in detecting prey or predators, schooling, and orienting to objects in the water. "Keep quiet or you'll scare away the fish" is good advice when you are fishing.

Life in the Water

Fish have been on Earth for more than 400 million years. Today there are more than 36,000 species identified worldwide. They are divided into three major classes: (1) jawless fish, which include lampreys; (2) **cartilaginous** fish, which include sharks and rays; and (3) bony fish, the largest group.

In Texas, there are over 250 species of freshwater fish and 1,500 species of saltwater fish in the Gulf of Mexico. Each fish species has its own way of surviving. The diversity of fish and their many adaptations allow them to live in a variety of environments. They thrive in the cypress swamps of the Sabine River and Caddo Lake, in complete darkness in the Hill Country's many caves, and in the brackish bays and estuaries along Texas' 367-mile-long coastline.

Different species of aquatic life are adapted to play different roles in the aquatic environment. Just like fish, many species of invertebrates, amphibians, reptiles, mammals, and birds have adaptations that allow them to survive in aquatic ecosystems.

Aquatic Invertebrates

There are many kinds of aquatic **invertebrates**, ranging from giant squid and clams, to squirmy insect **larvae** that live in the mud, to tiny free-swimming zooplankton. Invertebrates make up much of the food larger aquatic organisms eat. In addition to being part of the food web, aquatic invertebrates help break down organic matter. They are excellent indicators of the health of an aquatic habitat. Species can range from pollution tolerant to extremely sensitive to pollution.

Most aquatic invertebrates require oxygen to live, which they get in various ways, depending on the species. For example, some aquatic insects have small gills along the body or tail, others have little holes along their body to absorb dissolved oxygen from the water, some use breathing tubes they raise from the water into the air, and others can carry a small bubble of air underwater with them from the surface.

Mollusks are species that live in freshwater and saltwater and are extremely varied in form. Some mollusks, including clams, oysters, scallops, mussels, and snails, have hard shells (fig. 4.3). Others, like the octopus, have soft bodies and tend to live in cavities for protection; squid, which also have soft bodies, are free-swimming species.

Aquatic insects are found only in freshwater and the shallow brackish waters of estuaries and bays. These invertebrates have at least two phases of life. Insects change form through a process called **metamorphosis**. For insects such as many mayflies, stoneflies, dragonflies, and damselflies, the

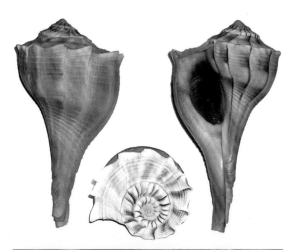

4.3. The lightning whelk (*top*), shown with a strand of egg capsules (*bottom*) from which tiny whelk juveniles emerge, is an aquatic invertebrate and the state shell of Texas. Along with clams and oysters, the lightening whelk is a mollusk. Photographs by Michael Barruel distributed under a CC-BY 2.0 license (*top*) and Hans Hillewaert distributed under a CC BY-SA 3.0 0 license (*bottom*).

Figure 4.4. Mayfly nymph (*top*) and adult (*bottom*). This insect is an aquatic invertebrate. Photographs courtesy of Valerie Bugh, larvalbug.com.

larval or **nymph** phase is usually spent entirely in water, but the adult phase may be spent in water, on land, or in the air. Many adult aquatic insects have large wings that allow them to fly (fig. 4.4). Although no aquatic insects can live entirely in saltwater, many inhabit brackish waters, such as saltwater mosquitoes.

Crustaceans also live in both freshwater and saltwater. They have an **exoskeleton**, which is an outer covering that supports and protects the animal's body. Familiar crustaceans are crayfish in freshwater and shrimp and lobster in saltwater.

Plankton are tiny invertebrates and photosynthetic organisms that are carried about by flowing waters or ocean currents. Animal plankton, or **zooplankton**, are made up of tiny crustaceans and even tinier animals called rotifers. Some species can be as large as your fingernail, but most are so small you need a microscope to see them. While many are capable of very quick movement on their own, they are so small that the movements they make are tiny. Zooplankton are important food for small fish (fig. 4.5).

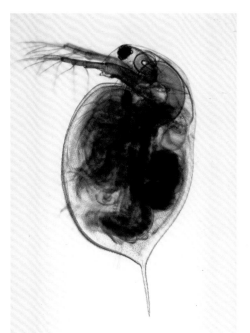

Figure 4.5. Daphnia, a crustacean zooplankton, is an aquatic invertebrate. Photograph courtesy of Public Library of Science; PLoS Biol 3(7):e253.doi:10.1371/journal.pbi0.0030253.g001.

Aquatic Plants and Algae

Many species of plants have special adaptations for living submerged in water or at the water's surface. Aquatic and wetland plants do not belong to any one particular plant family. They come from several land plant families and have acquired similar special adaptations to allow for life in water. The most common adaptation is large air spaces or channels running through the leaves, stems, and roots. These air channels allow an exchange of gases between the parts of the plant that remain submerged in water, such as the root, and the parts that reach above the water's surface. Other adaptations are floating leaves and leaves divided into many deep, narrow segments. Aquatic plants fully adapted for a life in water can grow only in water or in soil that is saturated with water (fig. 4.6).

A few aquatic plants are able to survive in brackish water or saltwater, but only in shallow areas. Several species of seagrass are found in Texas bays and estuaries. Seagrass beds are important to the estuarine food chain. Decomposing seagrass leaves provide nutrients for small shrimp, crabs, and fish. Seagrass leaves also provide protective cover for these small animals. Waterfowl, such as redhead ducks, also feed on seagrass leaves and roots.

Seaweed in the ocean and estuaries is often confused with aquatic plants, but seaweed is multicellular **algae**, not a vascular plant. There are many forms of algae, some of which are so small they can be seen only with a microscope. These algae are part of the phytoplankton in freshwater and saltwater.

Figure 4.6. Many aquatic plant species, such as these water lilies and duckweed, inhabit the waters of Caddo Lake in East Texas. Photograph courtesy of Texas Parks and Wildlife Department.

Phytoplankton

Phytoplankton are photosynthetic organisms in the plankton. They are sometimes called microalgae because they are so small and are a form of algae. While most are far too small to be seen without a microscope, they can grow together in very large groups. In fact, the dark green color of many ponds is due to the presence of very large numbers of tiny algae. Phytoplankton contain **chlorophyll** and require sunlight to live and grow. Many phytoplankton species are buoyant and remain near the water's surface during the day where sunlight penetrates the water. Phytoplankton also require nutrients such as nitrates, phosphates, and sulfur, which they convert into proteins, fats, and carbohydrates. There are many kinds of phytoplankton. The dinoflagellates use a whiplike tail, called a flagellum, to move about in the water. Their bodies are covered with a shell. The **diatoms** are a type of algae encased within a cell wall of interlocking parts made of silica and can form colonies (fig. 4.7).

Phytoplankton may be eaten by small fish, but most are eaten by zooplankton. Phytoplankton play an even bigger role in aquatic systems than just as food for aquatic animals. Phytoplankton are important **primary producers** in the aquatic food chain. They obtain energy through the process of photosynthesis by living in the sun-lit surface layer (termed the **euphotic zone**; *eu*

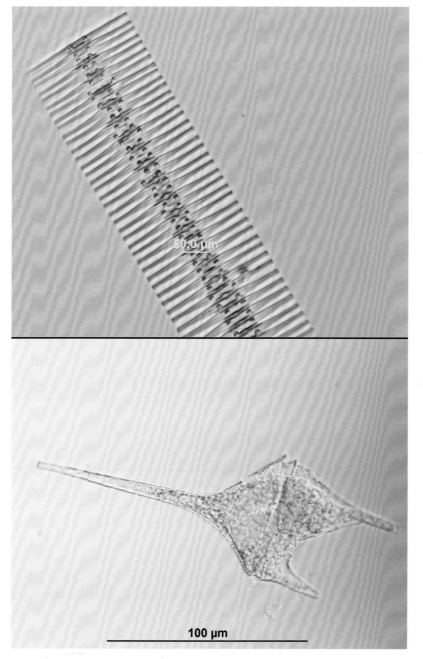

Figure 4.7. Phytoplankton: diatom (*top*) and dinoflagellate (*bottom*). Photographs courtesy of Texas Parks and Wildlife Department.

means "true," and *photic* means "light") of the Earth's oceans, lakes, and other bodies of water. Through this process, they produce organic compounds from carbon dioxide dissolved in the water. They also produce oxygen that can either remain dissolved in the water or can rise into the atmosphere. Phytoplankton account for half of all photosynthesis that occurs on Earth.

ADAPTATION TO AQUATIC ECOSYSTEMS FROM HEADWATERS TO OCEAN

Photograph courtesy of The Meadows Center for Water and the Environment.

Texas blind salamander

The Texas blind salamander lives in underwater caves in the Edwards Aquifer and springs that feed water to the headwaters and streams in the upper reaches of the Guadalupe River. Because it lives its life totally in the dark, it has no eyes and has smooth, completely colorless (unpigmented) skin. It gets oxygen from small gills that look much like feathers attached along its side just behind its head. The salamanders are about 3 to 4 inches in length.

Photograph courtesy of Thomas, Bonner, and Whiteside, *Freshwater Fishes of Texas* (Texas A&M University Press, 2007).

Orangethroat darter

Only about 1.5 inches long, orangethroat darters live in headwaters and streams. They remain near the bottom or among rocks and other hiding places where they dart about in fast-flowing water and in the small pools formed in streams. They are well adapted to hugging the bottom when streams flood and flows increase because most darter species lack a swim bladder, which is present in most freshwater fish.

Illustration courtesy of Texas Parks and Wildlife Department.

Guadalupe bass

Guadalupe bass live in rivers and streams of the Brazos, Guadalupe, Colorado, Nueces, and San Antonio River basins. They are adapted to life in small streams and may grow to about 1.5 feet long. They eat insects, crayfish, and small fish. The Guadalupe bass is the official Texas state freshwater fish and lives only in Texas. It is currently listed as a threatened species due to habitat degradation and interbreeding with stocked smallmouth bass and is the focus of restoration efforts by the Texas Parks and Wildlife Department.

Photograph courtesy of Texas Parks and Wildlife Department.

Blue catfish

Blue catfish are opportunistic predators and scavengers along the bottom of reservoirs or large rivers. They feed on fish, crustaceans, insects, and mollusks—dead or alive. They will even eat some plant matter, but as they grow, they feed more and more on live fish. They depend on barbels or "whiskers" with many taste buds and a good sense of smell to guide them to food, even in dark, muddy waters. In fact, catfish do not have scales but have chemoreceptors all over their skin, so they can taste food even before taking it into their mouths.

LIVING IN WATER

Photograph courtesy of The Meadows Center for Water and the Environment.

River cooter

This turtle is often seen basking in the sun on riverbanks or logs. It can grow to 1 foot long and eats aquatic plants, grasses, and algae. Large webbed feet make the river cooter an excellent swimmer, capable of negotiating moderately strong river currents.

Photograph courtesy of Texas Parks and Wildlife Department.

Largemouth bass

Largemouth bass live in reservoirs, wetlands, and large rivers where waters are slow moving. They are predators, meaning that they eat other animals, which are called prey. Their large mouths enable them to catch frogs, fish, crayfish, and other animals, including smaller largemouth bass. Their broad fins and strong, heavy bodies allow them to go in any direction (even backward). They can move quickly for short distances to capture their food. The largemouth bass is the most popular freshwater fish among people who enjoy fishing in Texas.

Photograph courtesy of Texas Parks and Wildlife Department.

American alligator

American alligators are reptiles that can be found in freshwater wetlands and brackish waters of Texas. They can grow to more than 12 feet long. Their large, powerful tails allow them to move swiftly in the water to catch prey, such as fish, turtles, small mammals, birds, and even other alligators. They often can be seen floating just below the water's surface with only their eyes and snout poking out from the water. Placement of their eyes and nostrils is an adaptation to life in water that allows them to breathe air and see around them while staying submerged and hidden.

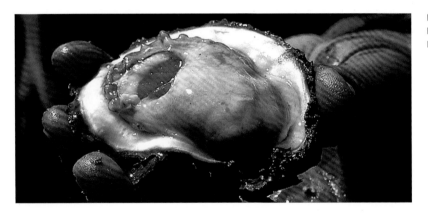

Photograph courtesy of Harte Resesarch Institute for Gulf of Mexico Studies.

Eastern oyster

Oysters live in estuaries and bays, feeding on plankton by using their gills to filter tiny food particles from the water. At two weeks of age, oyster larvae, each about the size of a grain of pepper, settle onto objects to which they attach with cementlike glue. At this stage, they are called spat. They begin a complete metamorphosis process to build their shell and spend the rest of their life in this one place. Oysters have been found attached to bricks, boats, cans, tires, bottles, crabs, and even turtles, but they usually attach to other oysters. When a large number of oysters join together, they form an oyster reef. More than 300 different marine aquatic species have been found to use oyster reefs for habitat, refuge, and food.

LIVING IN WATER

Photographs courtesy of Texas Parks and Wildlife Department.

Mallards (*top*) and Roseate spoonbill (*bottom*)

Waterfowl include ducks, geese, and swans. They have webbed feet adapted for swimming and can float on the water's surface as well as dive for short distances underwater. Other waterbirds have long, thin legs that allow them to wade. Many also have long beaks that they dip down into the water to pick food items off the bottom of shallow streams, ponds, and estuaries.

CHAPTER 4

Photograph courtesy of Texas Parks and Wildlife Department.

Red drum

Red drum, also called redfish, live when they are young in shallow, brackish waters about 1 to 4 feet deep along the edges of bays with submerged vegetation such as seagrass. They are also commonly found around oyster reefs and in soft mud along jetties and pier pilings. They sometimes are in water so shallow that their backs are exposed while swimming. They can live in freshwater, but when they are about three to six years old, they move out into the deeper waters of the Gulf of Mexico, where they can live entirely in saltwater.

Photograph by Jennifer Idol, The Underwater Designer.

Kemp's ridley sea turtle

One of five sea turtle species found in the Gulf waters of Texas, the Kemp's ridley sea turtle is an endangered species. These turtles can grow to 32 inches long and weigh as much as 100 pounds. Sea turtles' feet are formed as flippers, allowing them to swim through the water. These turtles lay eggs along beaches in Texas. Once hatched, the young turtles make their way across the beach and into the Gulf of Mexico, where they spend their entire lives. Although they live in the ocean, turtles do not have gills, so they must come to the surface to breathe.

Photograph by Allison Henry, National Oceanic and Atmospheric Administration–NEFSC/Department of Commerce.

Bottlenose dolphin

Bottlenose dolphins are mammals that breathe air and live their entire lives in water. They are often seen in Texas bays and passes to the Gulf of Mexico. They travel alone or in small groups called pods. They feed on fish and mollusks, with mullet their main food in Texas. They find food by using sound waves in a process called echolocation, in which the dolphin makes sounds and listens to the echoes of those sounds that return from bouncing off objects near them.

AQUATIC SCIENCE CAREER

Fisheries Biologist

A fisheries biologist usually works both in the field and in a laboratory and office. Outdoors work takes fisheries biologists to streams, lakes, estuaries, and the ocean. They use specialized gear to catch fish to investigate many aspects of fish biology, such as fish population numbers, migration patterns, and the food fish are eating. They also gather data on habitat important to fish species. Sometimes fisheries biologists get information from anglers and commercial fishermen to determine impacts of fishing on fish populations and economic impact of fishing on local economies. Many fisheries biologists specialize by studying a specific fish species or group of species, or they may study fish in specific lakes, rivers, or oceans.

Laboratory work can involve analyzing the data taken from the field and writing reports on what they find. They use computers to develop models of how fish interact with the environment or how fishing is affecting the number or size of fish in a population. Sometimes they take samples of fish from polluted waters to determine if they are free of contaminants and safe to eat. Fisheries biologists work for state and federal fisheries agencies, private conservation groups, consulting firms, and universities. Fisheries biologists usually have a master's or doctorate degree.

Photograph courtesy of Texas Parks and Wildlife Department.

chapter 5

■ ■ ■ ■ ■ ■ ■

From Sun to Sunfish

Questions to Consider

- What are some of the basic survival needs of all living things?
- What is a population? What is a community?
- What is habitat? Why is it important? Why must organisms compete for resources?
- What is a niche? Why is it important?
- What is carrying capacity?
- What is the source of energy for aquatic communities? How does energy circulate among organisms in an aquatic community?
- What is a food chain? What is a food web? What is an energy pyramid? What is a trophic level?
- How do predator and prey species keep species populations in balance in aquatic communities?
- What is natural selection?
- What are invasive species? Why are they a problem?

CHALLENGE QUESTIONS

In what ways might food webs, food chains, and predator-prey relationships be different in a pond, an estuary, and the Gulf of Mexico? What happens when one piece is altered or removed?

What do you need to survive? You need air to breathe, clean water to drink, food to eat, and shelter to protect you from the elements. These are basic survival needs for humans. For aquatic organisms to survive, they must be able to meet all of their survival needs just as you do. In addition, even though it does not influence the survival of any one individual, reproduction must occur for a species to continue to exist.

A group of individuals of the same species living in the same place at the same time is called a **population**. A group of the same kind of algae, a group of the same species of mussels, and a group of red-eared slider turtles are all examples of populations you might find living in a river. Several species living in the same place make up a **community**. The algae, mussels, and turtles—along with all the other aquatic organisms living in or around the river—interact with each other and make up a river community.

The physical environment that a species needs to survive is its **habitat**, but habitat is more than a place. Habitat is the shelter a species uses to escape predators and the elements. It is the space a species needs for reproducing and for containing all the individuals in the population. And habitat includes the food a species needs for hunting, gathering, or producing. It includes all the things a species needs and most or all of what it prefers.

Good habitat provides everything a species needs to survive over time, such as appropriate temperature, adequate dissolved oxygen, cover (such as logs and aquatic vegetation), or a particular bottom substrate. For example, channel catfish are native to Texas and are well adapted to the state's warm climate and aquatic habitats.

Survival of any population of aquatic organisms, or of a species over time, depends on having suitable habitat. If conditions in the habitat change, populations must either be able to adapt to the new habitat or move to another habitat in which they can survive. Aquatic plants and animals that have very specific habitat needs are called specialists. Specialists are the most susceptible to being harmed by changes to their habitat. These species are often unable to move or find another suitable habitat and therefore have difficulty surviving over time. Generalists are able to survive in a wide range of habitat conditions. Rare and endangered species are often specialists that have been unable to adjust well to changes in their habitat.

Finding a Niche

Within a community every species has a particular **niche**. A species' niche defines how the species fits into its environment. It includes its way of getting food, the habitat it needs, and the role it performs in the community. The diverse aquatic environments in Texas provide many different niches. For example, gizzard shad and largemouth bass both make their homes in large reservoirs. Gizzard shad, however, feed mostly by grazing on plankton, which are microscopic plants and animals. If necessary, they can also eat small aquatic insects. A largemouth bass hunts and eats other fish and crayfish. However, when bass are very young and small, they also seek plankton and small aquatic insect larvae. These two fish species live in the same reservoirs, but they eat different things. They do not **compete** for the same food, except

Figure 5.1. Flathead catfish can live in the same reservoir with gizzard shad because they do not compete for the same niche. Gizzard shad eat plankton and spawn over gravel and grass. They broadcast their eggs, which sink and adhere to any underwater substrate. When they are small, flathead catfish eat invertebrates, such as worms, insects, and crayfish. But once the flatheads grow large enough, they begin to prey on live fish. They spawn in sheltered areas on the lake bottom, such as cavities in logs, undercut banks, and rocks. Flathead catfish males guard the eggs. Once hatched, the fry remain on the nest for about a week, still guarded by the male. Illustrations courtesy of Missouri Department of Conservation and Texas Parks and Wildlife Department; modified by Rudolph Rosen.

for a short period of time when bass are very young. There are sufficient food resources for both species to survive over time because the bass and the shad occupy different niches in the same environment by eating different foods.

Largemouth bass and gizzard shad also have very different places and ways to breed. Largemouth bass spawning begins in the spring when water temperatures reach about 60 °F (15 °C). Depending on location in Texas, this can be as early as February or as late as May. Male bass build a circular nest about twice as far across as the bass is long. Nests are usually in water about two to eight feet deep. Once a female largemouth bass lays its eggs in the nest (between 2,000 and 43,000 eggs!), she is chased away by the male, who then guards the eggs. The young hatch in about 5 to 10 days. The newly hatched fish, or **fry**, remain in a group or "school" near the nest and under the male's watch for several days after hatching before swimming off on their own.

Gizzard shad also spawn in shallow water in spring when water temperature reaches about 60 °F (15 °C), but that is where the similarity with bass ends. To reproduce, gizzard shad males and females school together, releasing **milt** and eggs simultaneously near the surface, where the eggs are fertilized. Once fertilized, the eggs become sticky. The eggs are carried by water currents, and they adhere to underwater objects as they slowly sink to the bottom. A single female can release as many as 400,000 eggs. These hatch in about four days. Immediately after hatching, the fry form schools and swim away. Gizzard shad do not make nests or have any parental involvement. As the young gizzard shad hatch and mature, they become food for the young largemouth bass once they switch from a plankton to a fish diet.

This story is similar for other species where largemouth bass and gizzard shad live, such as the flathead catfish. Different species may have similar or even overlapping habitats, but no two species can occupy exactly the same niche in the same community for long without competition adversely affecting one or the other (fig. 5.1).

Competition and Survival

Living organisms have the capacity to produce populations of an unlimited size if they have unlimited food and other necessary resources, but this is a situation that never exists for very long. When there is not enough of something to go around, individuals must compete for whatever becomes scarce. If it is something necessary for survival or desirable to the individuals of any one species, some or perhaps all individuals can be adversely affected.

Individual bluegills in a pond compete with one another for food. Populations of species within a community may compete against one another as well. Bluegills in a pond compete with green sunfish, since both species are similar and feed on the same prey. This spells trouble for both species when food is scarce.

The amount, extent, or quality of biotic (living) and abiotic (nonliving) resources needed by a species in any one place determines the environment's **carrying capacity**. Carrying capacity is the maximum number of individuals in a particular population that an environment can support. When there are more resources than a particular population can use, the population is below carrying capacity for that particular environment. When this happens, individuals can continue to grow and reproduce.

When there are more individuals in a population than the environment can support, the population is above carrying capacity. Populations usually do not stay above carrying capacity for long. Once a population exceeds the habitat's carrying capacity, individuals may starve, get sick, or be forced to move to a place that can support them. Some examples of resource limits in aquatic habitats are the availability of food and cover (fig. 5.2).

Figure 5.2. When a population exceeds the carrying capacity of its environment, starvation and disease may result. Illustration courtesy of Missouri Department of Conservation.

Solar-Powered Systems

Aquatic communities, like almost all natural systems, run on sunlight. Plants and algae capture the sun's energy. Then by the process of **photosynthesis**, plants and algae use energy from the sun's light to make food and oxygen for themselves out of basic environmental resources like carbon dioxide, minerals, and water. Plants and algae are called **producers** because they produce their own food.

Unlike plants, animals cannot make their own food. To survive, they must be **consumers**, or eat other living things. Animals that eat plants are called **primary consumers**, or **herbivores**. A pond snail is a primary consumer because it eats algae and other aquatic plants. Secondary consumers eat primary consumers. Bluegills are secondary consumers—they eat insects, crustaceans, and small fish. Secondary consumers can be **carnivores** or **omnivores**. Carnivores eat other animals. Omnivores eat both plants and animals. **Parasites** such as leeches get their energy directly from feeding off other living organisms, but they usually do not kill the organism in the process. **Scavengers** such as crayfish eat the organic material of dead plants and animals.

Decomposers such as bacteria and fungi also feed on nonliving organic matter often called detritus. In the process they break it down into simple molecules that plants can use. Scavengers and decomposers play a vital role in recycling the energy and materials from the flesh of dead organisms directly back into the ecosystem for producers and other consumers to use.

Food Chains, Food Webs, and Energy Pyramids

Energy moving from producers to primary consumers to secondary consumers and so on is called a **food chain**. Food chains describe what eats what. For example, zooplankton may eat phytoplankton, gizzard shad may eat zooplankton, and largemouth bass may eat gizzard shad. Such descriptions of energy flow through a community of plants and animals give only a simplified version, because most animals have many sources of food at any point in time. Their food source may vary by life stage, size, and season of the year. Each food source may feed many different kinds of animals. To illustrate this, **food web** diagrams show how different food chains are interconnected (fig. 5.3). Taking out any link in a food chain may upset the balance of the whole food web.

An **energy pyramid** is another way to look at feeding relationships and energy flow through an ecosystem. Energy comes from the sun. The largest number of species and the greatest amount of available energy are in the producers, which take the sun's energy and produce food. This forms the base of the energy pyramid. As we move up the pyramid, energy is transferred up to organisms at higher levels. Energy is also transferred out of the pyramid at each step, and fewer species and individuals can be supported. The smallest number of species and least amount of energy are in the carnivores and top

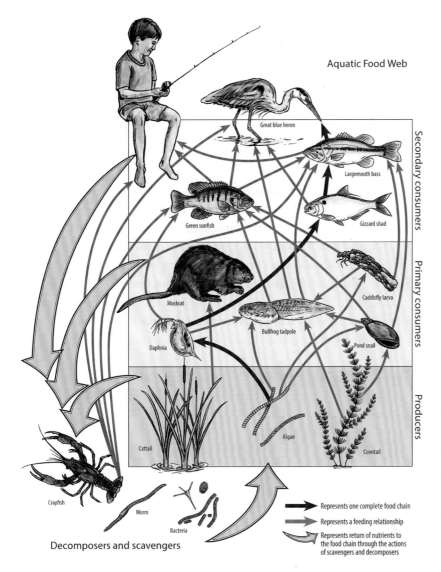

Figure 5.3. Food webs are interconnected food chains that display the feeding relationships within a community. A food chain shows energy flow through the community (shown by the arrows in the picture) as a result of what eats what. Where the bases of the arrows converge, two or more species eat the same thing and competition between species may occur. Illustration courtesy of Missouri Department of Conservation.

predators. The pyramid shape not only shows us what eats what but also how much energy is available at each level in the environment. Some energy is lost at each step away from the source of energy, the sun.

The groups of organisms that occupy the same position in a food chain, such as the producers and consumers, are called **trophic levels** in the energy pyramid. Trophic levels describe the steps in the energy pyramid and organisms' roles at each level. Only a little of the sun's energy passes from one trophic level to the next (fig. 5.4). Animals lose energy doing tasks such as hunting and keeping their bodies warm.

An example of lost energy happens when a whooping crane eats a blue crab (fig. 5.5). The crane gets any energy that is in the crab's body when it gets eaten, but the crane does not get the energy the crab spent that day walking on the bottom and swimming around the wetland before it was eaten. That energy is lost. An energy pyramid illustrates this lost energy by showing each higher trophic level having a smaller volume of energy than the one below it.

Most of the available food energy in a food chain is lost as it moves up each trophic level. For example, it takes about 3,200 pounds of microscopic plants to produce 410 pounds of microscopic animals. Those 410 pounds of microscopic animals can feed 58 pounds of crayfish, snails, mussels, clams, and aquatic insects. Those animals may in turn be eaten by up to 8 pounds of bluegill. Eating 8 pounds of bluegill will allow a largemouth bass to grow by about 1 pound.

An environment can support only a certain amount of life at each step of the energy pyramid. The higher up the energy pyramid an animal feeds, the fewer of this kind of animal the environment can support. Most energy pyramids can continue for only four or five trophic levels and can support only a few top-level consumers. In Texas' reservoirs, for example, largemouth bass and striped bass often occupy the top level of the food chain. These top-level predators are fewer in number than smaller fish species, invertebrates, or aquatic plants. In large aquatic ecosystems, such as the Gulf of Mexico, most of the energy in the system comes solely from within the system. In smaller systems such as freshwater streams, much of the energy used in the ecosystem comes from plant and animal matter that falls into the water. This comes from the stream banks, riparian vegetation, and material that is blown or thrown into the stream.

Figure 5.4. Beginning with the sun, energy flows through the aquatic community. All organisms capture a portion of that energy and change it to a form they can use. Only a portion of energy from each level of the energy pyramid is transferred upward from one level to the next. The process of energy transfer is inefficient. Considerable energy at one level of the community is used up by hunting for food, generating body heat, or carrying out other life processes before it can be transferred to the next level. Illustration courtesy of Missouri Department of Conservation.

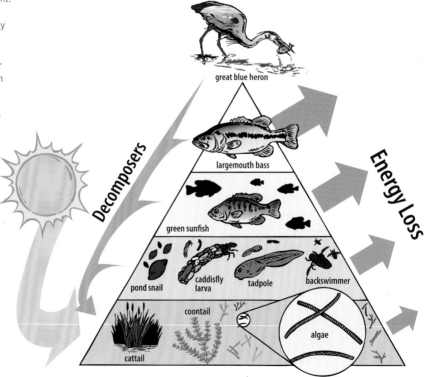

Figure 5.5. Blue crab is the favored food of the whooping crane, one of the most endangered species in the world. Many whooping cranes spend the winter on the Texas coast near Rockport at the Aransas National Wildlife Refuge. Photograph courtesy of Texas Parks and Wildlife Department.

Eat and Be Eaten

Aquatic species must compete for aquatic resources to survive. One adaptation that reduces some of the stress of competition is for species to specialize in the way they get their food. For example, think of water in a column. Fish that feed on insects that fall in the water from vegetation along the shoreline or stream bank often eat near the surface. Fish that prey mostly on smaller fish may roam throughout the water column. Some fish, such as catfish, feed near the bottom. Feeding at different levels in the water column helps one fish species reduce competition with other species. Feeding specialization helps a species to fit into a particular niche.

Predation is a form of competition. Both predator and prey are competing against one another for survival. The **predator** wants to eat the prey, while the **prey** wants to keep from being eaten. This is a life-or-death struggle for both predator and prey. To complicate matters, a species may be both a predator and prey at the same time. Many fish species shift from being prey when young and small to being a predator when older and larger. Some species, such as bass, even prey on their own young. Predator-prey relationships develop naturally within a community. It is a balancing act, as the numbers of predators and prey vary over time. The process is essential to keeping the populations of organisms at about the carrying capacity of their habitat.

Here is an example of how changing the balance between predators and prey can affect an aquatic system. Largemouth bass are a predator, and bluegill are their prey. Catching and removing all the largemouth bass in a pond leads to an overabundance of bluegill, since bass are no longer eating bluegill. No longer being eaten by bass, the bluegill quickly overpopulate their habitat and exceed its carrying capacity. The now large number of bluegill quickly eat all of their available food, such as all the appropriately sized aquatic insects and zooplankton. Without the zooplankton grazing on the phytoplankton (algae), the algae quickly become overabundant. The pond then turns green and turbid with algae. The algae grow too rapidly and use up all the available plant nutrients in the pond water. Without nutrients, the overgrown algae die all at once. Decomposers such as bacteria then feed on the dead algae, which forms large amounts of detritus. Decomposition of the large amount of detritus uses up all the dissolved oxygen in the water. With little dissolved oxygen in the water, bluegill in the pond become overabundant, grow very slowly, and may even die if oxygen levels get too low. But channel catfish survive, because bluegill require more dissolved oxygen than do channel catfish. The lesson of this example is that by simply removing a single species from the pond, such as the predatory largemouth bass, the entire aquatic community can undergo profound change that affects many other species.

Survival of the Fittest

Competition between individuals of a species is the driving force behind **natural selection**, the process whereby some individuals live to reproduce in their environment while other individuals do not. Natural selection ensures that, in general, only the best-adapted individuals survive and reproduce. Natural selection happens as individuals of a species compete for resources, such as food, water, territory, and sunlight or for mating partners. Where individuals of a species vary in characteristics that provide advantages, these individuals "win" because they are the ones best suited for survival in their environment.

The tendency for the most-adapted organisms to survive and reproduce helps the population pass on useful traits and abilities to future generations. However, it is not always clear what will be advantageous at any point in time or in different places. For example, having the ability to withstand low dissolved oxygen is not an advantage if you are a fish that lives where there is always plenty of dissolved oxygen. As a result, certain characteristics of individuals of a species change over time as the species adapts to the environment in which it lives. All aquatic plants and animals, including fish, have adapted characteristics over millions of years that allow them to live in today's aquatic habitats.

Not Every Species Is Welcome

Although having biodiversity in an ecosystem is important, sometimes new species are introduced into an environment that do not belong there. These are species that naturally occur outside Texas or even outside North America. When a species of foreign origin is introduced into a new habitat, it is called an exotic or **nonindigenous species.** Sometimes these species do not survive because some of the resources they need are not available in the new habitat (such as food or shelter). But when nonindigenous species do have the resources they need to survive in the new habitat, they can become invasive. An **invasive species** will typically reproduce quickly, consume resources faster than native species do, and is not subject to control by native predators. Invasive exotic species can cause damage to native aquatic habitats and even drive some native species to extinction. Lionfish, native to the Indo-Pacific, are now thriving in the Gulf of Mexico and eating our native reef fish. Giant salvinia can now be found on Caddo Lake. This is an example of an invasive exotic plant that is causing problems for native species in Texas reservoirs.

INVASION OF THE ZEBRA MUSSELS

Nonnative invasive species threaten native aquatic ecosystems in Texas. A species is considered to be invasive if it outcompetes native species for habitat, food, or both. This competition makes it harder for the native species to survive. Over time, invasive species may adversely change native Texas aquatic communities. As a result, native species can become endangered and the invasive species can spread throughout the state.

The tiny zebra mussel shown here has grown in massive numbers, infesting Lake Texoma in North Texas (many zebra mussels are shown clinging to the shell of a common Texas river mussel). Zebra mussels can attach to any hard surface. They clog water intakes, cling to boat hulls, damage engines, and infest boat docks, ramps, and navigational buoys. Aquatic ecosystem managers are concerned that zebra mussels will spread to many other Texas waters because they can attach themselves to parts of boats or be present in water trapped in boats. When boats are moved from one water body to another, zebra mussels can be carried along and released into new waters. They can then wreak havoc in the new environments, negatively impacting fish and native mussel populations by outcompeting them for available plankton. The zebra mussels can even make beaches unusable because they can completely cover the bottom along the shoreline.

Photographs courtesy of Texas Parks and Wildlife Department.

AQUATIC SCIENCE CAREER

Aquatic Science Laboratory Technician

Technicians do many different things in aquatic science laboratories. They may test fish for contaminants, evaluate how well hatchery fish are digesting different kinds of food, measure the salinity in water, count the number of tiny aquatic animals in a water sample by using a microscope, or conduct many other scientific measurements and tests. They usually work under the direction of an aquatic biologist, chemist, or other senior-level scientist. Laboratory technicians have at least a bachelor's degree in a science, such as biology or chemistry.

Photograph courtesy of Texas Parks and Wildlife Department.

chapter 6

■ ■ ■ ■ ■ ■ ■

Texas Aquatic Ecosystems

Questions to Consider

- What is an ecosystem? What are some of the parts of an ecosystem?
- How do the parts of an ecosystem interact with one another?
- What kinds of aquatic ecosystems do we have in Texas? How are they alike or different from one another?
- What is biodiversity? Why is it important?
- How do humans impact aquatic ecosystems?
- How can we help conserve aquatic ecosystems?

CHALLENGE QUESTION

How is the diversity of species in Texas' aquatic ecosystems connected to the economic well-being of citizens of Texas?

An **ecosystem** is a complex web of relationships between living and nonliving things. The study of ecosystems is known as **ecology**. The **biotic** parts of an ecosystem are the living components, such as the communities of plants and animals, including humans. The **abiotic** parts are the nonliving components, including sunlight, air, water, temperature, and minerals. Each part of an ecosystem is connected to and depends on all the others parts. All the parts must interact in a balanced fashion to make the system work. Changes to any part of an ecosystem can affect many others, which in turn may affect many more. A healthy, balanced ecosystem provides for the needs of the communities of life that are part of the ecosystem.

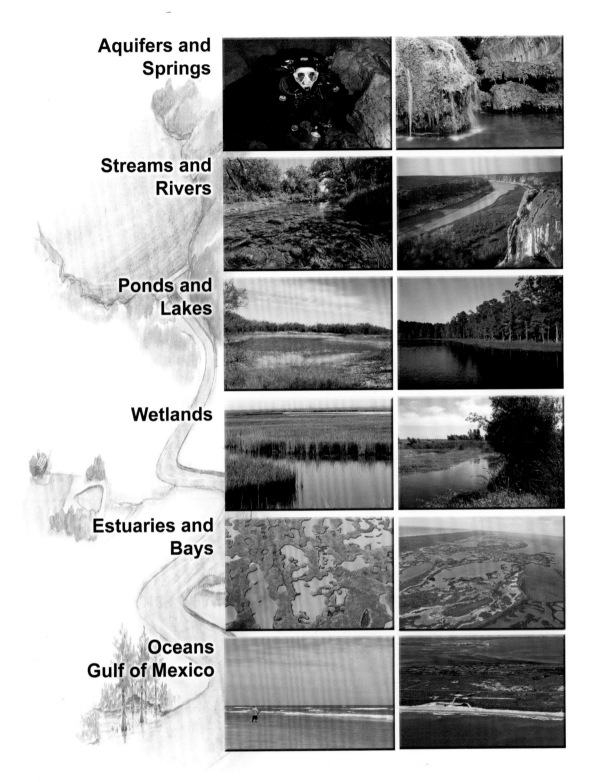

Figure 6.1. Aquatic ecosystems of Texas. Illustration courtesy of Missouri Department of Conservation; photograph by Jennifer Idol, The Underwater Designer (*top left*); all other photographs courtesy of Texas Parks and Wildlife Department; modified by Rudolph Rosen.

Kinds of Aquatic Ecosystems

Ecosystems organized around bodies of water are called **aquatic ecosystems**. Texas has six kinds of aquatic ecosystems (fig. 6.1):

- aquifers and springs
- rivers and streams
- lakes and ponds
- wetlands
- bays and estuaries
- ocean (Gulf of Mexico)

Aquifers and Springs

Aquifers are underground reservoirs and rivers of water that contain groundwater. Aquifers form where water seeps into the ground over time. Sometimes this water is ancient, having fallen on the land thousands of years ago.

Most of the land surface in Texas sits over aquifers, some large and some small. About half of the water we use in Texas is pumped from these aquifers by human-made wells drilled deep into the Earth. Aquifers need water seeping back into the ground to replenish the water that is pumped or drained out. These areas are called "recharge zones." A recharge zone is the area of land above an aquifer where surface water soaks back into the ground or travels through cracks or fractures between rocks deep into the Earth. Aquifers are very important to the economy of Texas and one of the reasons why people are able to live, ranch, and farm throughout the state.

Springs are the points where, because of the underground geology, groundwater travels to the surface and emerges from the ground. Springs can be a slow seep, or spring water can bubble up in pools and ponds. Springs sometimes form the headwaters of streams and rivers. Texas is home to over 3,000 springs, including some of the largest in the United States.

Rivers and Streams

Rivers and streams are flowing water with a measurable current. The current flows between the stream banks and over an underwater streambed or riverbed. The word *stream* can be used to describe all flowing natural waters. Rivers are just large streams. Streams are ever-changing systems that move and store water. They also move and store sediment and organic matter, such as leaves that fall into the water.

Texas has different kinds of streams that vary from one end of the state to the other. Streams have different sizes, shapes, lengths, flow rates, plants, animals, water quality, and streambed composition (for example, rocks or sand). Regardless of their size, shape, or location, all healthy Texas streams and rivers share a common feature: they are diverse ecosystems. The plants and animals living in them exist in balance with the processes that recycle

nutrients, or chemicals in the water that organisms need to grow. The healthiest streams are those that flow freely and have natural banks and streambeds.

Streams and rivers are important to people both ecologically and economically. They have been used from the earliest times by humans for travel and commerce. They carry the water humans need for life, and they often are used to carry away their wastes as well. Streams and rivers contain fish and attract wildlife that people use for food. They are the means by which freshwater is carried to the ocean, forming ecologically important wetlands and estuaries along the way. They also provide places for recreation important to the economy of Texas, including paddle sports, fishing, wildlife watching, and hunting.

Lakes and Ponds

Lakes and ponds, which are bodies of standing (not flowing) water, are among Texas' most well-known and popular aquatic ecosystems. You may be surprised to learn that Texas actually has only one natural lake (fig. 6.2).

The many lakes and ponds in Texas have been built by humans by placing dams across rivers or streams. These range from small ranch and farm ponds of less than the size of a football field to large lakes such as Lake Fork near Dallas and Lake Travis near Austin. The largest of these lakes **impounded** by dams are usually called **reservoirs**. There are more than 200 major reservoirs and more than 5,000 smaller ones in Texas.

Texas lakes and ponds hold water for use by people for drinking, production of electric power, recreation, and agriculture, such as for **irrigation** and for ranch animals. Boating, fishing, water skiing, and other outdoor experiences add billions of dollars to the state's economy each year as people buy equipment for these recreational opportunities and pay to travel and stay at outdoor locations.

Figure 6.2. Caddo Lake, located on the border between Texas and Louisiana, is the only naturally formed large lake in Texas. Although a natural logjam created the lake, today dams and reservoirs keep its waters under human control. The wetlands in Lake Caddo are so important that the lake has received international protection under the Convention on Wetlands of International Importance. Despite such protections, the lake's native aquatic life is currently threatened by giant salvinia, a fast-spreading invasive aquatic plant. This plant grows so fast it can double in size every two to four days, rapidly covering the water's surface with leaves and killing off life in the water below. Texas Parks and Wildlife Department biologists pictured here are sampling giant salvinia that is infesting the lake's waters. Photographs courtesy of Texas Parks and Wildlife Department.

Small ponds are formed by trapping water in valleys or other low spots in a watershed. Ponds are usually shallow enough that if the water is clear, sunshine can reach the bottom, allowing rooted plants to grow completely across it. A pond's water temperature changes with air temperature and is about the same from one end to the other and from top to bottom. There is little wave action, and the bottom is usually covered with mud. Lakes are bigger than ponds, so they are often too deep for light to reach the bottom and grow plants much beyond the shoreline. Lakes and ponds have much in common, but a lake's larger size and greater depth create differences in physical and chemical characteristics, including changes in dissolved oxygen and the temperature from top to bottom.

Wetlands

Wetlands are the "in-between" places, where water meets land. These are lands covered with shallow water at least part of the year. They can be present along the edges of rivers or lakes as the transitional zone between uplands and deep water, or wetlands can be individual bodies isolated or connected to other water bodies by groundwater. Think of wetlands as giant sponges laid out on the ground. When it rains, these are places where water collects and is held or where water slowly drains away over time. Wetlands can be big or small, full of tiny floating plants or massive trees. They are found from mountaintop to estuary. They are coastal shorelines, marshes, stream banks, and swamps. Life gathers around wetlands, and wetlands give life.

Wetlands are among the most productive ecosystems in the world. Texas has many large and ecologically important coastal wetlands. These kinds of wetlands are situated in estuaries and bays. Up to 90% of Texas' coastal saltwater fish species depend on wetlands for food, spawning, and places where their young can hatch and grow.

Yet wetlands have a bad reputation among some people. There are those who think these shallow waters are nothing more than stinky, bug-infested wastelands. Some people even think wetlands should be drained and used for other purposes. The truth is that healthy wetlands are very important to us. They help maintain water quality, recharge aquifers, reduce flooding, and provide habitat, and they are great places to go paddling, hunting, fishing, and wildlife watching.

Wetlands are dependent on the presence of water for all or part of the time. Thus, wetlands that do not have water in them year-round can sometimes be difficult to recognize and protect. Texas has lost over half its wetlands, many as a result of human alterations, draining, and filling. Many of the wetlands that are left have been partially filled in, polluted, or altered to the point they no longer function naturally. Taking care of the wetlands that are left and restoring them are among the biggest challenges facing natural resource managers today.

Bays and Estuaries

The terms "bay" and "estuary" are often used interchangeably. Bays are waters partially enclosed by land that open directly to the ocean. A bay's water can be brackish or saltwater. Estuaries are partially enclosed waters on the coast created where one or more rivers flow in, mixing freshwater with the ocean's saltwater. This produces **brackish water**.

Freshwater moves from land to the ocean in various ways—as flowing rivers and streams, as runoff from land near the coast, and as spring flow from aquifers. At the ocean's edge these freshwater inflows mix with saltwater to create brackish-water estuaries. This **freshwater inflow** delivers essential nutrients and sediments along with the freshwater.

Salinity is a measure of how much salt is in the water and is affected by how much freshwater reaches the coast. Salinity plays a critical role in the health of fish and other coastal plants and wildlife. Too little freshwater allows the estuaries to become too salty for many plants, fish, and wildlife to survive.

This dynamic mixing of freshwater and saltwater produces nutrient-rich, dark-colored, turbid waters. These waters feed estuary and bay habitats upon which 90% of all the commercially and recreationally important fish and shellfish of the Gulf of Mexico depend. The mud and sand bottoms of the estuaries and bays in Texas are dominated by extensive seagrass beds and benthic communities, including numerous oyster reefs. This essential habitat is being lost. About 60% of Texas' shoreline is eroding away, and almost half of the original coastal wetlands are gone.

Barrier islands run along the entire coast. These long, narrow islands block direct flow of freshwater to the ocean, creating productive estuaries and helping protect bays and shorelines during hurricanes and storm surges. Barrier islands host fascinating, yet sometimes fragile, ecosystems.

Ocean (Gulf of Mexico)

The Gulf of Mexico is the ninth-largest body of water in the world. It is shaped like a giant wide-brimmed bowl with the edge full of shallow bays and estuaries (fig. 6.3). Starting from the coastline and moving out into the ocean, the Gulf has wide and shallow shelves that gradually slope into the deeper Gulf waters. The floor of the Gulf is mostly a vast expanse of undulating soft mud bottom. The freshwaters that flow into the Gulf greatly affect the health of the aquatic life there. For example, the water flowing from the Mississippi River along the Louisiana coast has created a hypoxic zone, often called the dead zone. This is a vast area deficient in dissolved oxygen where many organisms become stressed or cannot survive. This environmental threat comes from excessive nutrients and wastes carried into the Gulf by the Mississippi River, which collects water from 41% of the continental United States.

All of these factors combine to support one of the most productive, yet ecologically threatened, bodies of water in the world. The Gulf is a place of

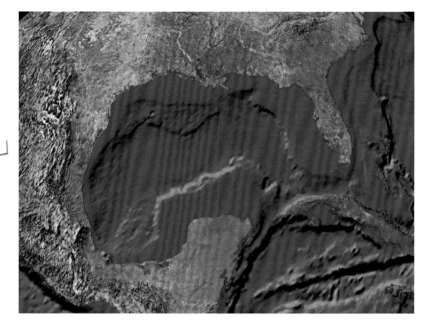

Figure 6.3 The Gulf of Mexico is an international aquatic resource. It is bordered on three sides by the United States, by the eastern coastline of Mexico, and by Cuba in the southeast. Map courtesy of Harte Research Institute for Gulf of Mexico Studies.

incredible biodiversity, with over 15,000 species calling it home. Commercial fisheries annually catch more than 1.5 billion pounds of **seafood**. Shrimp are the predominant species caught for food for people. Gulf shrimp and oysters account for 70% of all the shrimp and oysters that go into grocery stores and restaurants across the United States. Recreational fishing is also important, providing employment and tourist dollars in coastal communities (fig. 6.4).

Over the last 90 years, the Gulf and US coastline has changed dramatically. Fifty percent of our coastal wetlands are gone, up to 60% of the seagrass beds have been lost, and over 50% of our oyster reefs have disappeared. In Texas, reduced freshwater flowing into estuaries and bays reduces the amount of certain kinds of habitat needed by many of the Gulf's most important species to people and the aquatic ecosystem.

Texas' Aquatic Systems Are All Connected

Life in healthy aquatic ecosystems is constantly progressing toward a state of balance, but balance does not mean a lack of change. Ecosystems are always changing. Change comes in response to natural or human-caused events. For example, heavy rains can force a river to change course, leaving the old channel and forming a new one. A human activity such as straightening a stream speeds up erosion and cuts out curves that shelter fish and other aquatic life. Changes may destroy habitat for some species and create habitat for others.

Whether changes are good or bad depends on how they affect the ecosystem's **biodiversity**. This term refers to the variety and number of different species and populations. The more closely the biodiversity in an ecosystem matches that of a completely natural system, the healthier, more sustainable,

and better balanced it is. Some human activities that can reduce aquatic biodiversity are draining a wetland, damming a river, or pumping so much water out of an aquifer that springs no longer flow. These activities destroy habitat, which is the main cause of species decline. Therefore, protecting and restoring a wide variety of aquatic habitats help keep species from becoming endangered or extinct.

Figure 6.4. Forty-five percent of all the people who go fishing in saltwater in the United States fish in the Gulf of Mexico. Illustration courtesy of Texas Department of Transportation.

AQUATIC SCIENCE CAREER

Environmental and Endangered Species Protection Worker

There are many different kinds of jobs where you can work to protect the environment and help endangered species. Environmental protection work includes environmental law, politics, activism, journalism, fundraising, education, and science. You can find work in citizen environmental organizations, at federal and state agencies, and at universities conducting research. You can help build environmentally sustainable ways of life for the future. The level and kind of education required depend on the type of work. For example, work in environmental law requires a bachelor's degree and a law degree, while work as an activist may require only a high school degree.

Photograph courtesy of Texas Parks and Wildlife Department.

chapter 7

Aquifers
& Springs

Questions to Consider

- What is an aquifer? What is groundwater?
- What are headwaters?
- How are aquifers similar? How do they differ?
- How do aquifers recharge?
- What is a playa lake? What role does it play in Texas?
- What is a spring?
- How have springs influenced Texas history?
- What kinds of aquatic ecosystems exist in groundwater? What adaptations enable aquatic life to exist underground?
- How can we help conserve groundwater?

CHALLENGE QUESTIONS

How is your life connected to aquifers? Which aquifer provides groundwater where you live? How is your groundwater being used? Is it being conserved, or is it being depleted?

An **aquifer** is an underground layer of **permeable** rock or sand that collects, holds, and conducts water. The material acts like an underground sponge, allowing water to flow very slowly through it. Water in the aquifer is called groundwater. Many aquifers are like reservoirs because they store water useful to humans and aquatic ecosystems.

Groundwater may naturally emerge from the aquifer as springs. When an aquifer's water table reaches the surface, groundwater can be released as surface water into rivers and lakes. Wetlands may form where groundwater

Figure 7.1. Major aquifers of Texas. Map courtesy of Texas Water Development Board.

reaches ground level. If groundwater drops below the surface, these wetlands will become dry and no longer be wetlands.

Aquifers vary in size, from narrow to wide, and may be hundreds of feet thick. They may span one to two counties or may stretch across thousands of square miles and several states. They are **recharged**, or refilled, when precipitation falling on the land seeps into the ground. Aquifers give rise to Texas streams and rivers where springs form headwaters. People also drill wells into the aquifer to pump groundwater to the surface to use for drinking or irrigation.

Aquifers

Most of the land surface of Texas lies above aquifers. There are 9 major aquifers (fig. 7.1) and 21 minor aquifers (fig. 7.2) in Texas. A major aquifer

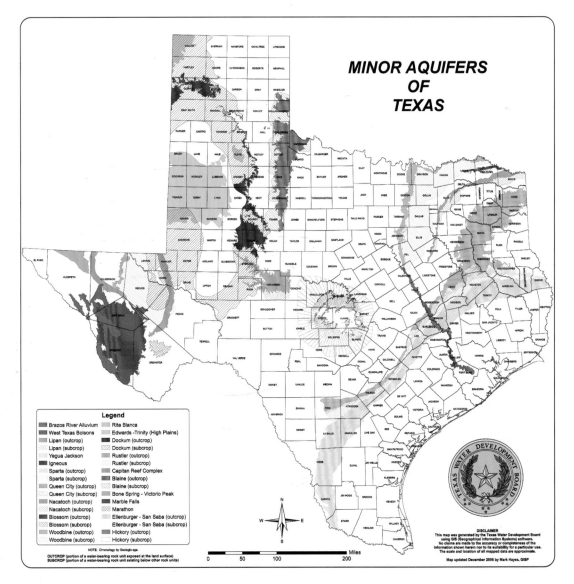

Figure 7.2. Minor aquifers of Texas. Map courtesy of Texas Water Development Board.

contains large amounts of water spread across a large area. Minor aquifers contain smaller amounts of water spread over large areas, or larger amounts of water spread over small areas.

Aquifers are an important source of water for humans, supplying about 60% of the water we use. Most of the water pumped from aquifers goes to agriculture for irrigation of food crops. Over 80% of the irrigation water used in Texas comes from one aquifer, the Ogallala. The largest aquifer in the United States, the Ogallala stretches from South Dakota southward into Texas and underlies much of the Texas High Plains region. This aquifer's thickness averages 95 feet, although it can be over 800 feet thick in some places.

The groundwater in some aquifers can be ancient. The Ogallala Aquifer was formed about 2 to 6 million years ago. Groundwater can also be very new.

Figure 7.3. San Solomon Springs is the largest artesian spring formed by geologic faults in the Balmorhea area, located in the foothills of the Davis Mountains in West Texas. The spring now feeds directly into the largest spring-fed swimming pool in the world, located in Balmorhea State Park. After leaving the swimming pool, spring waters flow through San Solomon ciénega. This desert wetland contains two endangered desert fishes: the Pecos gambusia and the Comanche Springs pupfish. Pecos gambusia have a very limited distribution, and Comanche Springs pupfish are found only in the waters of the Balmorhea springs. Shown to the right is a school of Mexican tetra, a species that is abundant in the pool. Photograph courtesy of Texas Parks and Wildlife Department (*left*); photograph by Jennifer Idol, The Underwater Designer (*right*).

Water that falls as rain and enters the Edwards Aquifer near San Marcos can be found emerging from springs a few days or even hours later. Parts of the Edwards Aquifer are like a giant underwater cave system. Groundwater can flow like a river through large openings in the underground **limestone** and **marble** rock that forms this aquifer.

Aquifers in Texas

Texas has three kinds of aquifers. **Unconfined aquifers** are directly connected to the surface and have water levels dependent on relatively constant recharge. Groundwater flows to the surface whenever the aquifer's upper saturated layer, the water table, rises to the level of the land's surface. Perhaps the best-known unconfined aquifer is the Ogallala. In many places this aquifer is near the surface, and recharge depends on water that collects at the surface in wetlands. As much as 95% of the recharge water in Texas comes from **playa lakes**. A playa lake is a naturally occurring wetland (averaging about 17 acres) formed when rain fills small depressions in the prairie. There are about 20,000 playa lakes in the High Plains of Texas.

Confined aquifers are saturated layers of **pervious** rock bounded above and below by largely **impervious** rock, which water cannot pass through. The aquifer's placement between impervious rock layers can "squeeze" the groundwater, putting it under pressure. A confined aquifer containing water under pressure is called an **artesian aquifer**. Artesian flow feeds many of Texas' famous springs, including San Solomon Springs in West Texas (fig. 7.3). Artesian pressure can be so great that if a well is drilled into the aquifer, groundwater can be pushed all the way to the surface without the need for pumps.

Karst aquifers are found in limestone and marble rock. Over long periods of time, limestone and marble can be dissolved by water, which can result in large holes, channels, and even large underground caverns, lakes, and streams. One of the most famous aquifers in the world, the Edwards Aquifer, is a confined karst aquifer flowing through limestone in Texas. There are many well-known locations where water from this aquifer flows to the surface, creating large springs. In some places cave entrances open directly on the surface and lead deep into the aquifer. Underwater divers have explored some of the Edwards Aquifer's underwater caves and flowing rivers (fig. 7.4).

Springs

Springs are the places where aquifer water flows to the surface. A pool and often a stream are formed by springs. Texas is home to more than 3,000 springs (fig. 7.5). Springs can be cool or so hot that the water steams and almost boils when it reaches the surface.

Springs often form along **faults**. These are places where an earthquake has cracked and split open the earth, exposing the aquifer's water-bearing rock. One such fault in Texas is the Balcones fault, which runs from approximately the southwestern part of the state near Del Rio to the north-central region near Waco along Interstate 35. For a distance of about 300 miles, this fault has exposed the Edwards Aquifer, creating many prominent springs.

Aquifer and Spring Ecosystems

Most aquifers do not support aquatic ecosystems as we normally think about them. There is no light underground, so no photosynthesis by plants can occur. There may be no dissolved oxygen. There are some organisms that can live in such dark, nutrient-poor anaerobic conditions. **Microorganisms**, primarily **bacteria**, **protozoa**, and other **unicellular** life, are found in aquifers. Some kinds of bacteria in groundwater are useful to humans. An example is **denitrifying bacteria**, which remove nitrates. Excess nitrates are a water pollutant that may seep into the aquifer, particularly in agricultural regions where fertilizers or manure containing nitrates are spread on soils to enhance growth of food crops.

There is one kind of aquifer that can support a more complex underground aquatic ecosystem, although the life here is very different from that we see in lakes, rivers, or the ocean. The karst aquifers' caves and underground lakes and rivers can support entire ecosystems that include invertebrates, fish, and amphibians. But these species are unlike any we are used to seeing aboveground. The aquifer has no sunlight and therefore no green plants or algae with chlorophyll taking the sun's energy and converting it to food. Without these primary producers, these aquatic systems do not have a lot of nutrients or food available. Available food is constantly recycled among the organisms, with only occasional additions from the outside. These underground

SPRINGS GAVE LIFE TO CENTRAL TEXAS TOWNS

Springs have played a large role in the economic development of Texas. Many cities were built near springs that provided water for Native Americans and early settlers.

Salado Springs

Salado was built around Salado Springs, the twelfth-largest spring in Texas. Salado Springs is located near the Stagecoach Inn on the east side of Interstate 35.

Barton Springs

Austin, the Texas state capital, was built where it is because of the consistent and abundant flow of water from Barton Springs, the fifth-largest spring in the state. Today the springs are more commonly known as Barton Springs Pool, which is located in Zilker Park and is open for swimming year-round.

San Marcos Springs

San Marcos was built around San Marcos Springs, the second largest in Texas. These springs were home to Native Americans long before Europeans arrived and are thought to be the longest continually inhabited site in North America. For many years it was the location of Aquarena Springs theme park, featuring glass-bottom boat rides so visitors could view the underwater life. Today the springs are part of the campus of Texas State University, where visitors can still take glass-bottom boat rides and learn about the spring ecosystem.

Comal Springs

New Braunfels was built at the largest of Texas' springs, Comal Springs. These springs are also the largest in the United States west of the Mississippi River. They are located at Landa Park and form the headwaters of the Comal River. Just downstream from the park is Schlitterbahn, where hundreds of thousands of Texans enjoy the Comal River's waters each summer.

San Antonio and San Pedro Springs

San Antonio was built around San Antonio and San Pedro Springs. San Antonio Springs flow out of the aquifer on the property of the Incarnate Word Sisters north of downtown San Antonio in a place called the Headwaters Sanctuary. The waters then flow through Brackenridge Park into downtown San Antonio. San Pedro Springs emerge in San Pedro Park, the second-oldest public park in the United States. Spanish missionaries located here in the early 1700s. San Antonio is the largest city in North America wholly dependent on groundwater. As a result, the city's future depends on maintaining the sustainability of the Edwards Aquifer, which feeds the springs.

Leona Springs

Uvalde was built near the Leona River, which is fed water from Leona Springs. This large group of springs now has reduced flows.

Las Moras Springs

Fort Clark, now called Brackettville, was built around Las Moras Springs. These springs supplied the military fort with water from 1852 until just after World War II. The springs now feed into a public swimming pool.

San Felipe Springs

Del Rio was built on Texas' fourth-largest springs, the San Felipe Springs. These are a group of 10 or more springs that extend for over a mile along San Felipe Creek on the grounds of the San Felipe Country Club and several ranches to the north of Highway 90.

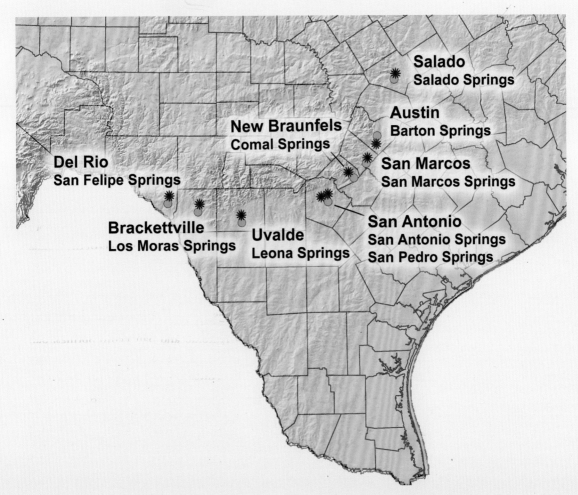

Figure 7.4. Jacob's Well is an artesian spring of the Edwards Aquifer that flows from one of the most extensive underwater cave systems in Texas. The spring itself is a 12-foot opening in the bed of Cypress Creek, a few miles north of Wimberley. The mouth of the cave serves as a popular swimming hole, but from the opening, Jacob's Well descends vertically for about 30 feet, continuing from there to a series of large chambers separated by narrow passageways descending to a depth of at least 120 feet. Divers have explored this cave and its aquatic ecosystem. The spring feeds Cypress Creek and Blue Hole, the popular swimming hole in Wimberley, before the water flows into the Blanco River about five miles downstream. Photographs by Jennifer Idol, The Underwater Designer.

CHAPTER 7

Springs

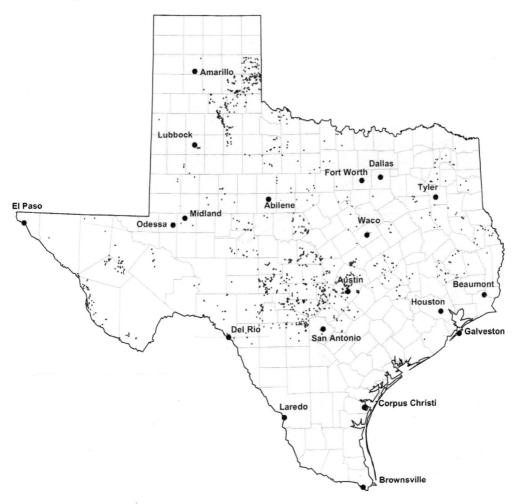

Figure 7.5. Springs of Texas. Map courtesy of Estaville and Earl, *Texas Water Atlas* (Texas A&M University Press, 2008).

ecosystems have a very low carrying capacity. They can support only a few individuals of any one species, and these individuals do not grow very large.

The lack of sunlight has another consequence. The single most amazing adaptation of invertebrates, fish, and amphibians to the dark underground aquatic ecosystems is an absence of eyes. Without light there is no need for eyes (fig. 7.6). Predators have adapted ways to find and catch prey in the dark, and prey have adapted ways to escape. Sensory adaptations such as antennae, **chemoreceptors**, and touch receptors are usually highly developed in underground-dwelling species.

These species also often have a very low metabolism, allowing them to live on very little food. This adaptation is aided by the constant temperatures of aquifer waters, unlike those of surface water, which can vary quickly and widely. Groundwater species live in a very stable and predictable environment.

Figure 7.6. The Texas blind salamander is adapted for living in the water-filled caves of the Edwards Aquifer. It has no eyes, only two small black dots under the skin, yet the blind salamander is an active predator. It hunts food by sensing water-pressure waves created by prey in the still, underground waters where it lives. Tiny snails, shrimp, and other aquatic invertebrates make up its diet. It has little skin pigment, is white in color, and has red external gills used to get dissolved oxygen from the water. It is unknown how many Texas blind salamanders exist. Photograph by Sam Massey.

As in aquatic ecosystems aboveground, there can be overlaps with other ecosystems. This means species in the aquifer may not be completely isolated from life on the surface.

The land surface above karst aquifers is an integral part of the habitat of animals living in the underground areas. Holes in the limestone and marble of these aquifers often extend to the surface. Jacob's Well, near Wimberley, is a good example. Here, a large opening in the streambed of Cypress Creek is actually a water-filled cave extending deep into the aquifer. There are also caves open to the surface that lead to the aquifer.

Because plants cannot grow in darkness, the cave and associated underwater ecosystem are dependent on plant and animal materials being washed into the cave from the outside. Food in a cave can also come from organisms such as bats, mice, and crickets that take shelter in caves. They can become food (prey) for cave dwellers or leave "food" behind when they leave. For example, bat and mouse feces dropped on a cave floor provides nutrients fungi need to grow. Fungi are eaten by several species of insects that may wander in and out of the cave. These insects reproduce rapidly, move about the cave, and become prey for predatory invertebrates that live their entire lives in the cave. These invertebrates fall into the water and are swept deep into the aquifer, where they can be eaten by species such as the Texas blind salamander. The salamander is eaten by the toothless blindcat, a catfish that lives more than 1,000 feet below the Earth's surface.

The aquifer ecosystem extends beyond the aquifer itself where groundwater emerges into the spring. Here the groundwater mixes with surface water in the spring, stream, rivers, and lakes. However, it is in the springs formed by the aquifer's emerging waters where the unique underground ecosystem truly extends to the surface. We occasionally get a rare glimpse of life in the

CHAPTER 7

underground ecosystem when an invertebrate, fish or, salamander from the Edwards Aquifer gets swept out at a spring. You may see one of these odd creatures, or one of the predators that likes to eat them, when you tour the San Marcos Springs in a glass-bottom boat.

Although many aquifers do not contain aquatic life, most major springs in Texas do. Some even contain species found nowhere else. The Edwards Aquifer ecosystem and its springs contain more than 60 species of plants and animals that live nowhere else in the world. Species of salamanders, fish, amphipods, beetles, spiders, and others have evolved in isolated habitats within the aquifer and springs. Many of these live in the dry caves above the water table, and others live in the many springs fed by the aquifer. Barton Springs, located in Austin, is the only place where the Barton Springs and Austin blind salamanders live. Fountain darters live only in the headwaters of the San Marcos and Comal Rivers. Texas wild rice lives only in the San Marcos Springs and River immediately downstream of the springs.

Because springs are exposed to sunlight, aquatic plants and algae play a role in providing nutrients. A relatively constant temperature, at least near spring openings, allows species to adapt to a highly specialized lifestyle. These species often cannot exist very far from the spring, including in the stream and river the spring creates. Water temperatures in particular may limit a species adapted to a spring ecosystem from wandering beyond the vicinity of the spring. Cool spring waters may quickly warm when exposed to hot weather. As a result, spring ecosystems tend to be small, allowing only relatively small populations of any one species to survive.

Like species adapted for a life completely underground, species in spring ecosystems may be greatly affected by even small changes to their habitat. For species in a spring, lowered spring flow due to drought or groundwater withdrawals may reduce habitat and create significant stress. Invasive species may quickly overwhelm and outcompete native spring species. Many of Texas' stream ecosystems have been impacted by both flow reductions and invasive species.

Humans have made the most changes to springs. Many of Texas' largest springs have even been turned into swimming pools by placing a dam across water flowing from the spring. These include Barton Springs Pool in Austin and the largest spring-fed swimming pool in the world, located at Balmorhea State Park in West Texas. At such places, the native spring ecosystems no longer exist.

Using Groundwater Wisely

Groundwater in an aquifer is like money in your bank. Each time you put money in your bank, the amount of money you have grows. When you withdraw money to buy something, your bank account goes down and stays down until you add more money. Aquifers grow when nature puts water in through

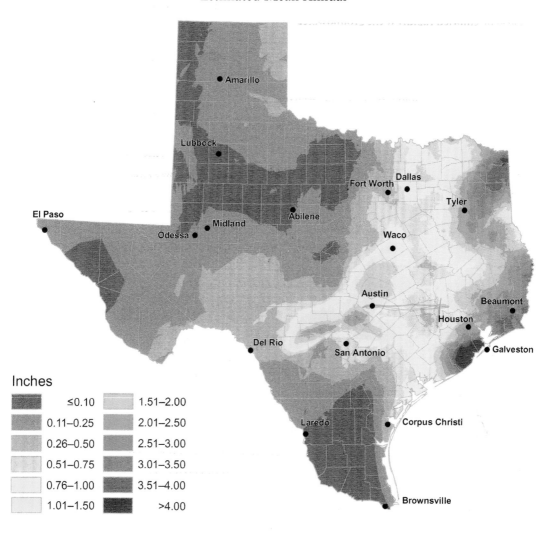

Figure 7.7. Groundwater recharge in Texas. Map courtesy of Estaville and Earl, Texas Water Atlas (Texas A&M University Press, 2008).

rainfall, snow, and other precipitation. Water leaves the aquifer naturally along streams, at springs, and through other openings. Humans greatly increase water withdrawals by drilling wells into the aquifer and pumping water out. Just as withdrawing money from your bank account depletes the money you have for future use, withdrawing groundwater without equal or greater deposits of rain depletes our groundwater.

The size of the aquifer and how full it is determine how much water it is holding at any point in time. The porosity of the aquifer determines the rate at which water can move through or into the aquifer. Some aquifers refill rapidly, so if there is sufficient rainfall in the aquifer area, the amount of time needed to recharge the aquifer may be short (fig. 7.7). The Edwards Aquifer

AQUATIC SCIENCE CAREER

Stream Ecologist

Freshwater stream and river ecologists study the animal and plant life in rivers and streams. They may conduct surveys of stream life, run statistical analysis, or do research studies. It is interesting to work from a boat, wade through streams, or even work in a laboratory using special tanks that imitate stream habitats. Ecologists can find work in state and federal agencies and environmental consulting firms. Ecologists usually have a master's or doctorate degree in aquatic biology or a related field.

Photograph courtesy of Texas Parks and Wildlife Department.

chapter 9

Lakes & Ponds

Questions to Consider

- Where are ponds and lakes in your community? What role do they play in your economy?
- How do temperature and oxygen levels in ponds change during each 24-hour period?
- How are lakes similar to ponds? How are they different?
- What kind of organism makes up the greatest amount of living material in a pond?
- Besides providing food, what other roles do plants have in lake and pond ecosystems?
- How are plants that live underwater similar to plants that live on land? How are they different?
- How do ponds change over time?

CHALLENGE QUESTION

What are the benefits and costs of building new reservoirs as a solution for the future water needs of Texas? Consider the economy and the environment in your discussion.

Texas had more than 1.2 million acres of freshwater lakes, ponds, and reservoirs in 2013. These bodies of **lentic water** (water that is not flowing) are among Texas' most well-known and popular aquatic ecosystems. Almost all of Texas' lakes and ponds were built by placing dams across streams or rivers (fig. 9.1), so they can also be called reservoirs. Usually only very large bodies of water are actually named "reservoir." All the rest of the water bodies are usually called a lake or pond, depending on size. These range from small ranch and farm ponds of less than an acre to large lakes containing millions of acre-feet of water, such as Lake Lewisville near Dallas, Lake Travis near Austin, and Lake Amistad on the Rio Grande.

Lakes, ponds, and <u>reservoirs</u> have been built to hold water for use by people for drinking, production of electric power to reduce flooding, use in agriculture such as for watering crops and ranch animals, and for recreation such as fishing and boating. This water is critical to the state's economy (fig. 9.2).

There are more than 200 major reservoirs and more than 5,000 smaller ones in Texas. The only naturally formed lake in Texas is Caddo Lake, created

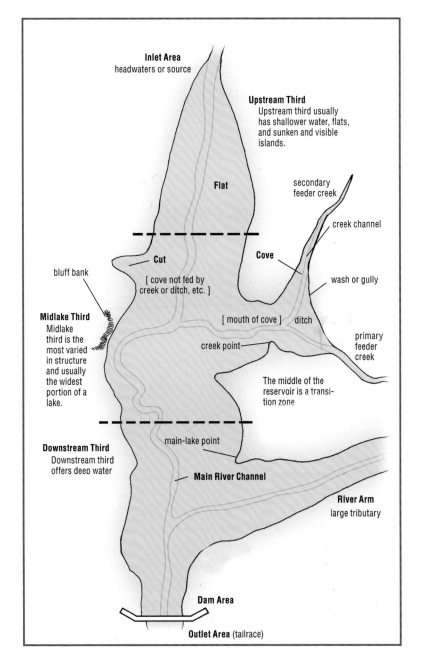

Figure 9.1. Almost all lakes and ponds in Texas were built by placing dams across streams or rivers. These can range in size from small farm ponds and stock tanks, to very large reservoirs where many people boat, water-ski, and fish. Illustration courtesy of In-Fisherman.

Figure 9.2. Fishing in Texas' freshwaters is worth over $2 billion to the economy of the state. Photograph courtesy of Texas Parks and Wildlife Department.

by a large logjam hundreds of years ago on the Red River. Although a natural logjam created the lake, today dams and reservoirs keep its waters under human control.

Ponds Are Small

Ponds are usually shallow enough that if the water is clear, sunlight can reach the bottom and plants can grow throughout the pond. Plant roots grow into the pond bottom and hold the soil, making the water even clearer and allowing more plants to grow at greater depths. This is important to the life of a pond because the plants produce much more than food. Being shallow allows a pond's water temperature to remain about the same everywhere in the pond. The pond's water temperature changes quickly as air temperature goes up or down.

Dissolved oxygen in a pond can also change fast. Underwater life depends on oxygen in the water. The main sources of dissolved oxygen are surface air and photosynthesis by plants and algae in the water. Oxygen is dissolved in water when water mixes with air at the water's surface. Waves and wind help mix air into the water. Water temperature and salinity can affect the amount of dissolved oxygen in water. Cold water can "hold" more dissolved oxygen than warm water, and freshwater holds more dissolved oxygen than saltwater.

Aquatic plants and phytoplankton are another source of oxygen. These organisms give off oxygen in the water as a by-product of photosynthesis. A pond's oxygen levels can vary widely over the course of a day because during the day air from the surface and oxygen from plants are constantly replenishing the dissolved oxygen consumed by animals and by **aerobic decomposition** of detritus and other decaying matter. Plants stop producing oxygen at night because photosynthesis requires sunlight. During the night, oxygen continues to be used by animals and in the process of aerobic decomposition. This results in many ponds having higher dissolved oxygen levels during the day than at night.

Lakes Are Bigger Than Ponds

While lakes and ponds have much in common, **lakes** are larger and deeper. This greater size and depth result in differences in dissolved oxygen levels, plant growth, and temperature. In a lake, the amount of oxygen dissolved in the water stays pretty even over a 24-hour period. Although a strong wind can ruffle up a pond's surface, on a lake it can whip up high waves. This mixes air into the water, helping increase dissolved oxygen levels.

In at least some places, the water in a lake is too deep for plants to grow on the bottom. Only in shallow areas, such as around the shore or islands, will there be enough sunlight reaching the bottom to allow plants to grow. This is the **littoral zone** (fig. 9.3). Lakes often include a transition zone containing a narrow band of wetlands extending out from shore. The ecology of the shallow shoreline zone is like pond ecology.

Toward the middle of the lake, away from shore, is the open-water zone called the **limnetic zone**. This part of the lake is too deep for sunlight to reach the bottom, so no plants grow on the bottom there. Despite not reaching bottom, sunlight still shines into the water in this zone to some depth. The part of the limnetic zone that gets sunlight is the **photic zone** (also called the euphotic zone). Some species of large and open-water fish spend much of their time in the photic zone. They may swim into the littoral zone now and then to feed or spawn, but these visits are only temporary. There is also a deep-water zone, the **aphotic zone**, located just below the photic zone. There is too little sunlight or oxygen reaching this zone for plants or algae to

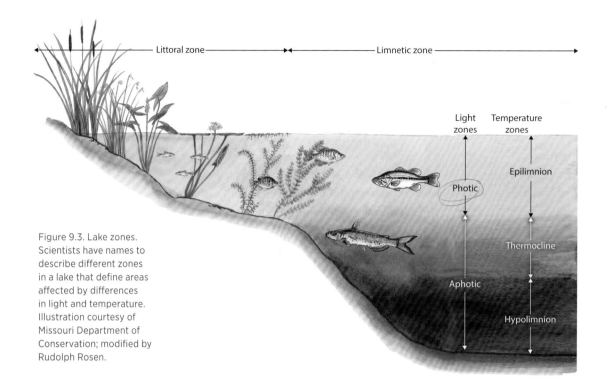

Figure 9.3. Lake zones. Scientists have names to describe different zones in a lake that define areas affected by differences in light and temperature. Illustration courtesy of Missouri Department of Conservation; modified by Rudolph Rosen.

grow. Dead organic matter sinks to the lake bottom, where bacteria and other decomposers such as some worms and larval stages of some insects break it down. This is called the **benthic zone**.

The overall temperature in a lake is fairly even from day to day, while surface temperatures can vary. Lake water in Texas is warmest at the surface and gets colder as the water becomes deeper. At the bottom of a lake, it is possible for water to become as cold as 39 °F (4 °C), the temperature where water is most dense, thus, heaviest.

Tiny Life Forms the Food Base of Big Ecosystems

Underwater plants and plankton make up most of the food base for entire pond and lake ecosystems. Plankton are tiny, mostly microscopic, free-floating organisms. There are two kinds of plankton. Phytoplankton are photosynthesizing aquatic organisms. They can grow in a lake's sunlit open-water photic zone, where rooted plants cannot. They are producers that release oxygen into the water and serve as food for many kinds of aquatic animals, including the second main type of plankton, the zooplankton. These are tiny, free-floating animals. Plankton have a high nutritional content and make up most of the living material in a pond. Zooplankton eat phytoplankton, and many species eat each other. They can be barely visible, complex organisms, such as water fleas, or they can be as small as single-cell protozoa.

Many invertebrates and small fish eat plankton. Some mollusks, such as clams and mussels, and fish can even filter plankton from the water by using specialized mouthparts and adaptations to gills. Organisms that feed this way are called **filter feeders**. Examples of filter-feeding fish in Texas lakes are gizzard shad and paddlefish. They spend most of their time in the open water where they can feed on plankton.

In the open water of a pond or lake it may seem that plankton have no defense to escape these filter feeders except by luck. Most zooplankton and some forms of phytoplankton have ways to move about. Despite being so tiny, many species can move very quickly. They also may hide among the plants growing underwater. But considering how fast and far even a small fish can move in comparison, plankton make easy prey for much of the aquatic life in lakes and ponds.

That is good for many of the fish species we know best. The top aquatic predators in lakes and ponds are fish we like to catch when we go fishing. Many species of fish get very large, but all fish depend in some way on the many tiny species at the bottom of the aquatic food chain. Species such as gizzard shad are grazers or filter feeders on plankton. Gizzard shad become prey for larger predator fish, such as largemouth bass and striped bass. Some fish, such as some catfish species, are scavengers and eat whatever they can find.

A Band of Life—Aquatic Plants

Like plants on land, aquatic plants spread out in beds or clumps and attract a variety of animal life. Some plants live entirely underwater (called submergent), while others have some of their parts sticking out of the water (called emergent). To survive, all plants need water, carbon dioxide, sunlight, and nutrients such as phosphorus and nitrogen. Water plants have special adaptations that help them thrive in their underwater environment. Waxy or slimy coatings protect them from drying out when water levels drop. Porous stems or leaves let them absorb minerals right from the water. Some have leaves that float on the surface, while others float entirely on the surface.

Sunlight can penetrate only so deeply into the water of a lake. Where sunlight reaches the bottom, rooted plants can grow. This littoral zone is often the area of the lake from the shoreline outward to the point where sunlight no longer reaches bottom. This is a narrow band or strip reaching out into the water from the shoreline and islands. These bands of plants hold the greatest biodiversity to be found in the lake ecosystem. They provide important habitat for many species of fish and invertebrates. While aquatic plants can become a nuisance if they overgrow their habitat, they play an important role in aquatic ecosystems and help maintain or improve water quality.

Nearest the shore, this area may look more like a wetland than a lake. It is a transition area between dry uplands and permanently deep areas of a lake or pond. Mud along the shoreline often contains tracks of all kinds of animals. Look carefully—maybe a deer or raccoon has been there.

The littoral zone's plant beds serve to shelter prey organisms from predators and are a food source for aquatic insects. In North America around 8,000 species of insects spend some or all of their lives in the water. Some of these insects feed directly on plants, while others feed on life found on the plants.

There are three main ways insects feed among a pond's plant life. **Scrapers-grazers** have special mouthparts they can use to remove algae growing on the surface of plants and solid objects; the mouthparts act like a sharp scraper blade. Collectors gather small bits of loose and decaying materials by using mouthparts or by brushing up the bits using fine hairs on their head or legs. Shredders use mouthparts designed to nibble off and then grind up pieces of live plants in the water or plant materials that fall into the water from plants growing along the water's edge. Omnivorous insects eat both plant and animal materials. Predator insects catch and feed on live animals, such as other invertebrates. Some aquatic insects are fast and large enough to catch and feed on small fish and tadpoles. Insects are not the only predators here. Fish and insect-eating land animals, in particular birds, seek out the insects in the littoral zone. They help complete the complex food web of predators and prey supported by plant growth in Texas' lakes and ponds.

Ponds Do Not Last Forever

As water runs downhill through the pond's watershed, it picks up small bits of soil and anything else that can be moved. This erosion deposits sediment in the pond, filling it with soil and other materials, so the pond becomes shallower. Decaying plant and animal material will also accumulate on the pond bottom, adding more materials and enriching the sediment with more nutrients. Plants thrive in the rich sediment, grow, and take up more space. In time the pond will become a wetland. As it fills even more, it becomes a wet meadow. This natural process is called **pond succession** (fig. 9.4).

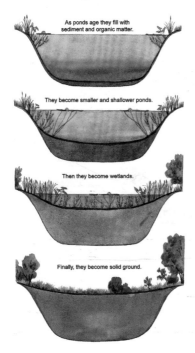

Figure 9.4. The natural process of pond succession fills in ponds over time. Materials from erosion in the watershed and plant and animal materials produced in the pond accumulate on the bottom, slowly filling in the pond. Illustration courtesy of Missouri Department of Conservation.

The surface water that fills the pond can also carry pollution into the pond. Soil and plant nutrients such as fertilizer washed down from homes and fields can be pollutants. Too much fertilizer unbalances the way in which nutrients naturally cycle through the pond's ecosystem. A common result is too much plant and algae growth. This process of excessive nutrient input and subsequent plant and algae growth is **eutrophication**. When excessive amounts of plants or algae die all at once, an unnaturally high level of anaerobic decomposition may result that can use up all the dissolved oxygen in the water. This can be so extreme that some fish in the pond may die from lack of oxygen.

Every body of water is only as clean and healthy as the water that drains into it off the land in the watershed, so good watershed management is important to keep a pond or lake healthy. Stopping excess erosion and runoff loaded with fertilizers, pesticides, or other pollutants is the most effective way to maintain a healthy pond or lake. Keeping a buffer of plant growth around the pond can also filter out pollutants and sediment before they reach the pond. A plant buffer can improve the health of a pond or lake and extend its life. In Texas, most ponds are built by people who want the benefits healthy ponds provide, so they often take actions to maintain ponds, including removing sediments.

Figure 9.5. Texas lakes are used for many kinds of fun activities, such as kayaking. Photograph courtesy of Texas Parks and Wildlife Department.

Figure 9.6. Each year in Texas, more than 20 million fish, including 21 different species, are stocked into Texas waters. Fishery managers conduct nearly 400 fish population surveys on about 150 of the state's largest lakes each year. This is done to ensure fish populations are healthy and to make fishing better in public fishing lakes. Photograph courtesy of Texas Parks and Wildlife Department.

Lake Tourism and Fishing Mean Dollars

Texans use lakes for drinking water, industry, and agriculture, but lakes are also where we go for recreation. Texas lakes provide water for fishing, boating, swimming, skiing, windsurfing, paddling, sailing, kiteboarding, and many other fun activities (fig. 9.5). When people visit a lake, they often buy gasoline, groceries, bait, and other items from local businesses. Lake-related recreation is important to the economies where lakes are located.

This is among the reasons why Texas fisheries managers work to create good fishing. Lakes and ponds contain lots of fish species popular with anglers. These species include bluegill, largemouth bass, catfish, crappie, striped bass, and white bass. While many fish in Texas' lakes and ponds are native to the state, some species have been introduced from elsewhere. Many fish are **stocked** each year in Texas lakes to provide better fishing opportunities (fig. 9.6). Some of the large lakes in Texas are famous for fishing, such as Lake Fork in East Texas, one of the country's best lakes in which to catch huge largemouth bass.

AQUATIC SCIENCE CAREER

Fish Hatchery Technician and Biologist

Freshwater fish hatchery biologists raise fish to be stocked into Texas lakes and rivers. Fish they raise may be used to bring back a species that has disappeared from a lake or to provide more fish for people to catch in a lake, stream, or bay. There are many kinds of jobs to do at a fish hatchery. Hatchery biologists work for state and federal fisheries agencies and private aquaculture farms. Workers who feed fish, maintain hatchery facilities, and do much of the fieldwork are often called technicians and have at least a high school degree. Many have a bachelor's degree. Biologists oversee the health and nutrition of fish, as well as ensure safe transport of fish to the places where the fish are stocked. Biologists usually have a master's degree in fishery biology or a related field.

Photograph courtesy of Texas Parks and Wildlife Department.

chapter 10

■ ■ ■ ■ ■ ■ ■

Wetlands

Questions to Consider

- What is a wetland? What factors must be present for a place to be considered a wetland?
- What are some of the different types of wetlands found in Texas? What are the differences and similarities between them?
- What are some examples of the special adaptations found in wetland plants and animals?
- How are wetlands important to fish, birds, and other wildlife?
- How do wetlands improve water quality?
- How do wetlands provide natural flood control?
- How do wetlands recharge aquifers?
- Why should we protect wetlands? How can we protect wetlands?

CHALLENGE QUESTIONS

What is the closest wetland to you? What type of wetland is it, and what value does it hold for your community? How does your community affect the wetland?

Wetlands are places where the land and water meet in a zone of transition—not dry, not a pond, but not land either. The key ingredient in a wetland is always water. All wetlands are wet for a major part of the plants' growing season. Some wetlands may have standing water, and others may just be muddy places. Some may even appear dry when it does not rain for a long time. But if you dig a hole, the hole will fill with water.

Smelly and Biologically Diverse

Because wetland soil is saturated, or covered with water, the tiny spaces between the soil's bits of dirt stay filled with water, leaving little or no room for air to get in. Mixed in with this water are tiny creatures that break down dead plant and animal matter, or **detritus**. Two types of decomposers help break down detritus. **Aerobic** bacteria play a role in the initial **decomposition** process. These bacteria require dissolved oxygen in the water, which the bacteria can quickly use up. Once this happens, **anaerobic** bacteria, which do not need oxygen, take over and do the majority of work in the wetland by feeding on detritus to break it down.

The decomposition process may produce sulfur-containing compounds that smell like rotten eggs. It may be stinky, but the smell tells us the wetland is healthy. The rich detritus nourishes a complex food web that supports a wide range of plants and animals. Next time you visit a wetland, grab a bit of the bottom and give it a good smell.

Wetland plants are specially adapted to live in the saturated soil. Many of the animal species found in wetlands can be completely aquatic, such as fish. Others require both wet and dry habitats for survival, such as beaver. Some species will use wetlands if available but otherwise can survive on land as long as they have some source of water to drink somewhere. The ability for so many species to use wetlands makes this kind of ecosystem one of the most biologically diverse on Earth. You can find more kinds of animals and plants in an acre of wetland than in almost any other kind of ecosystem (fig. 10.1).

Figure 10.1. Wetland ecosystems are among the most biologically diverse and productive in the world. Illustration courtesy of Texas Parks and Wildlife Department.

TEXAS COASTAL WETLANDS

This great diversity of life makes wetlands among the most productive ecosystems as well. Up to 90% of the saltwater and freshwater fish species in Texas depend on wetlands for food, spawning, and nursery grounds. Life gathers around wetlands, and wetlands give life.

Texas Wetlands

Wetlands in Texas can be divided into two broad categories:

1. *Freshwater wetlands* can form wherever shallow water collects on the land: river floodplains, bottomland hardwoods, **marshes**, **seeps**, springs, ponds, playa lakes, **sloughs**, oxbow lakes, and **swamps**; along some stream banks and lake areas; and places where the water table reaches the surface. Freshwater wetlands contain plant species adapted to life where water levels may go up and down. Many of these species can withstand periods when wetlands may become dry.
2. *Coastal wetlands* form where saltwater and freshwater mix together: coastal shorelines; shallow bays and inlets; and swamps, marshes, mudflats, and deltas of coastal lowlands and estuaries. Plant species must be able to survive changes in both salinity and water level, because these wetlands are often affected by changes in amount of freshwater inflow and tidal fluctuations in water level.

Within these categories, Texas has several kinds of wetland ecosystems. Where you are located in Texas will determine which kind of wetland you see (fig. 10.2).

Figure 10.2. Texas wetland regions and wetland types. Photographs and map courtesy of Texas Parks and Wildlife Department; modified by Rudolph Rosen.

HIGH PLAINS. Playa lake wetlands are found in the Texas High Plains. These wetlands form in shallow depressions in the land's surface. They are usually round and small and about 15 to 20 acres in size. The depressions naturally fill with water from rain or snow to form wetlands that are only about a foot deep. There are about 20,000 of these in Texas, and they play a major role in recharge of the Ogallala Aquifer. Playa lakes go through frequent wet and dry cycles. Riparian wetlands are also found on the High Plains. These are the wetlands that form along the edges of streams and rivers. They provide important food and cover for wildlife.

CENTRAL TEXAS. Spring-fed wetlands are an amazing geologic feature of Central Texas. These wetlands are filled by water that may have traveled great distances underground in an aquifer before flowing to the surface to form the wetland. Springs occur where there are faults, cracks, and other openings in the aquifer. Riparian wetlands are also found in Central Texas.

SOUTH TEXAS. Wetlands in South Texas are of two main types. **Sand sheet wetlands** are small, isolated depressions found in places where wind erodes away topsoil, exposing clay soils underneath. These depressions trap and hold water when it rains. **Resacas** are channels of the Rio Grande that have been cut off from the river. The channels fill with water and sediment, creating shallow wetlands and ponds. Sand sheet wetlands and resacas may be the only places wildlife can find fresh water in this very arid part of Texas. Both kinds of wetlands can become dry during periods of drought.

TRANS-PECOS. In the Trans-Pecos region, spring-fed wetlands found on the sides of mountains or in small mountain valleys are called mountain springs. These are small wetlands found in the Guadalupe Mountains, Chisos Mountains, and other rugged highland areas of West Texas. **Ciénegas** are another type of spring-fed wetland. These are small, isolated springs that occur on the desert floor in West Texas. Ciénegas and mountain springs provide water for plants and animals that could not otherwise survive in the desert.

EAST TEXAS. Bottomland hardwoods are the dominant wetlands in East Texas. These are known for the large trees that live in the water. They are also called forested wetlands. Unlike most of Texas, East Texas receives large amounts of rainfall. This rain floods streams and rivers, spilling water out onto the floodplain. The force of this flooding often reshapes the stream bottoms and floodplains, forming bottomland hardwood wetlands in heavily wooded areas. Floods can cut across stream and river curves, forming new channels where the water can flow. When this happens, **oxbow lakes** form in the "cut-off loop" of the stream channel (fig. 10.3). As the oxbow lake fills with sediment, it becomes a wetland. This is the same process that forms resacas on the Rio Grande. As these lakes fill in naturally with sediment, they transition from lakes to wetlands and then to dry land over many years.

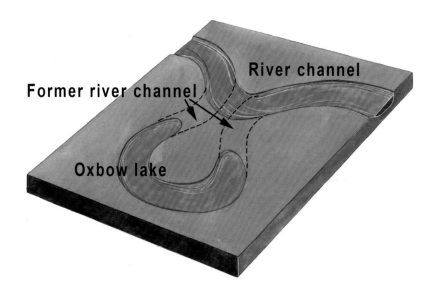

Figure 10.3. Oxbow lakes form when a flood or other natural process cuts off a loop of a river channel, isolating it from the main river. Illustration courtesy of Missouri Department of Conservation.

GULF COAST. Coastal wetlands are formed when saltwater and freshwater mix. Salinities vary widely as freshwater inflow to coastal areas goes up or down. There may be times the water is entirely fresh or times salinity can be greater than in ocean water. These wetlands can form in depressions in the land near the coast's bays and estuaries. Rains fill these depressions, and sometimes storms push saltwater into them. Coastal wetlands can also form in low-lying areas of land where rivers flow into estuaries and bays. Many coastal wetlands have water levels and salinities that fluctuate daily because they are subjected to tides. These are called tidal wetlands and are the tidal flats, bays, marshes, and bayous we see throughout the Texas coast.

Wetlands Are Special Ecosystems

Wetland plants are adapted to take advantage of every ray of sunlight yet live a life in water. They have ways to expose their leaves to the sun and avoid being shaded by other plants. They also have roots that can pull in water and still get air (fig. 10.4). Plants that grow in shallow water often have roots that tightly hold on to the soil so they can grow tall to reach sunlight. Cattails, rushes, sedges, and arrowheads do this very well. Sedges and rushes have air spaces inside their leaves to transport air and make the leaves buoyant (fig. 10.5). Some plants, such as the water lily, can grow in deeper water while still remaining anchored by roots in the soil because they have a very long, skinny stem (fig. 10.6). There are even plants that float around on the surface. The tiny duckweed has leaves with air spaces and grows in open water to avoid the shade of taller plants. Its roots are short and hang down in the nutrient-rich water. Coastal wetland plants, such as mangrove and salicornia, can live in water having a high salinity. These plants are called **halophytes**.

Animals that live in wetlands are special, too. Wetlands are home to many invertebrates, amphibians, reptiles, fish, birds, and mammals. Predators here

Figure 10.4. A cypress tree's "knees" are extensions of the tree's roots that project above the normal water level. They are thought to be an adaptation that allows these large trees to get more oxygen or gain a better foothold in the wetlands' constantly wet soils. Photograph courtesy of Texas Parks and Wildlife Department.

are adapted to find and catch prey in wet places. The whirligig beetle's eyes focus both above and below water level to help it find prey. The black-necked stilt's long legs and specially adapted feet allow it to walk on mud (fig. 10.7). Stilts can snatch fish and tadpoles from underwater with their long, slender beaks. The frog's long springlike legs and the turtle's shell help them escape predators. Ducks have spoonlike, flattened bills that make it easy for them to strain seeds and invertebrates from shallow water.

Even the fish in these shallow wet environments have special adaptations. With its upturned eyes and mouth, the mosquitofish can slurp down mosquito larvae on the surface. Some fish species spawn in the shallow, marshy

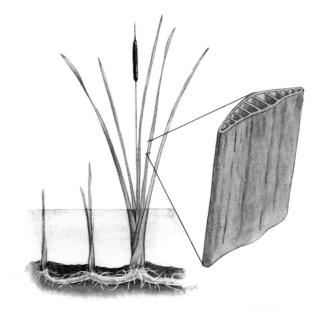

Figure 10.5. Cattail leaves have spaces that transport air to the roots. Illustration courtesy of Missouri Department of Conservation.

Figure 10.6. Lily pad leaves and flowers float on the surface, but long stems connect them to roots that hold firm to bottom wetland soil. Photograph courtesy of Texas Parks and Wildlife Department.

places along the shorelines of lakes and rivers. The small young fish hide from larger predators in the plant-filled, shallow-water wetlands. There is plenty of food for young fish in such places. They remain in this cover until they grow large enough to venture out into deeper water. Without the wetlands, these species would disappear, even though there may be plenty of deep water nearby where the adult fish are able to live. In fact, many freshwater fish and most of the important fish and invertebrates in the Gulf of Mexico are dependent on wetlands as a place for their young to feed and grow up safely.

Food and Lodging for Travelers

Wherever they may roam, **migratory** ducks, geese, shorebirds, and other waterbirds need wetland habitats. Migratory predator birds such as osprey, eagles, hawks, and owls also use wetlands, often as places where they go to feed on waterbirds.

The Texas coastal wetlands are especially important places for migratory **waterfowl** to spend the winter. Texas is at the southern end of the Central and Mississippi Flyways (fig. 10.8). These flyways are like highways in the sky that extend from one end of North America to the other. Migratory birds in Texas use these flyways to travel thousands of miles back and forth from our wetlands to nesting areas far to the north. As they travel, they stop and feed in wetlands along the way. Once they reach their northern breeding grounds, they lay their eggs, hatch, and raise their young. The birds return to Texas in early fall when northern areas begin to get freezing cold. After their long flight back, they spend the winter feeding in Texas' wetlands getting ready for the trip back north in early spring.

Figure 10.7. The black-necked stilt feeds in coastal mudflats, along lakeshores, and in other shallow wetland areas where it probes the mud and water for aquatic invertebrates—mainly insects, crustaceans, and mollusks—and small fish and tadpoles. Photograph courtesy of Texas Parks and Wildlife Department.

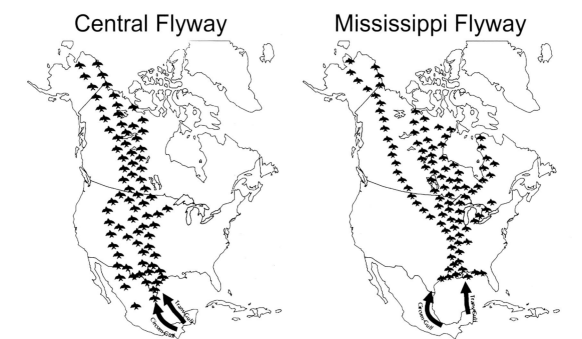

Figure 10.8. Texas' waterfowl migrate along the Central and Mississippi Flyways. Illustrations courtesy of Texas Parks and Wildlife Department.

As a result, wetlands in Texas are directly connected to wetlands in northern states, in Canada, and even as far away as the Arctic Circle. Healthy and productive wetlands are necessary at both ends of the birds' flight and all along the way. But wetlands in Texas play an especially important role for both wintering and migrating waterfowl.

Texas' prairie and coastal wetlands provide winter food and lodging for 90% of all ducks and 75% of all geese in the Central Flyway. For migrating birds, the 20,000 playa lakes in the Texas High Plains provide important stopover habitat. These wetlands supply plants, seeds, and invertebrates that migrating birds must eat to get enough energy to continue their flight north or south.

If the chain of healthy wetlands from north to south is ever broken, waterfowl like the blue-winged teal we see in Texas will be unable to survive and reproduce from one year to the next (fig. 10.9). Citizen conservation groups work together with Texas Parks and Wildlife Department and other state and federal wildlife agencies to protect and restore wetlands in Texas and other places where migratory birds travel. Migratory bird hunters buy special stamps to help pay for this wetland conservation.

Wetlands Improve Our Environment

Years ago, some people thought wetlands were just stinky, bug-infested wastelands. The truth is that wetlands are very important to us (even if they are sometimes stinky). You have already read how wetlands can serve as a place for wildlife to eat, rest, and even use as a nursery. Did you know that wetlands can also act as a filter, a sponge, and even a tourist attraction? Wetlands help

Figure 10.9. Blue-winged teal spend the cold winter months feeding and resting in Texas' coastal wetlands. Photograph courtesy of Texas Parks and Wildlife Department.

maintain water quality, recharge aquifers, reduce flood damage, and provide habitat and great places to go paddling, hunting, bird-watching, and fishing.

WETLANDS AS A FILTER. When it rains, water running off city streets, fertilized lawns, landfills, and some agricultural fields can pick up and carry high levels of contaminants that may be harmful to plants and animals. This runoff can flow into wetlands, but wetlands have the ability to absorb many of these kinds of pollutants, store them, break them down, and in some cases even use them as nutrients to grow wetland plants. Because of this amazing ability to clean up some pollutants, wetlands are being used to help treat wastewater from some cities and even livestock feedlots. Wetlands also improve water quality by cleansing or filtering runoff full of sediment that comes from higher in the watershed. This helps keep sediment out of our streams and rivers, which is considered their greatest pollutant.

WETLANDS AS A SPONGE. The flatness and lush plant growth in wetlands slow down the flow of rainwater so that water may gently trickle into nearby streams or seep into the aquifer, increasing the amount of groundwater. This ability of wetlands to recharge groundwater is especially important in times of drought and in arid parts of Texas where communities struggle to deal with declining water tables (fig. 10.10).

The ability of wetlands to slow the flow of water also helps reduce damage caused by floods. When the water from heavy rains reaches wetlands, the water is slowed and the wetlands act as giant sponges. They first absorb and hold the water and then release it slowly back into the watershed. Unfortunately, humans can disrupt this natural flood control by building levees along rivers, digging drainage ditches through wetlands, and **channelizing** streams. The unfortunate consequence of these actions increases damage from floods (fig. 10.11). Along the Texas coast, wetlands help protect shorelines and areas inland from flooding during huge storms, such as hurricanes.

Figure 10.10. Drought dries up Texas' wetlands, lakes, and rivers, harming entire aquatic ecosystems. Photograph courtesy of Texas Parks and Wildlife Department.

WETLANDS AS A TOURIST ATTRACTION. Wetlands help the economy and even attract tourists. Each year Texas fishermen catch about 30 million pounds of wetland-dependent shrimp with a value of $100 million. Texas is known around the world for great hunting and fishing. Without healthy wetlands the seafood, fish, birds, and animals we hunt and fish would simply not be here. Wetlands are also treasured by millions of photographers, boaters, hikers, and wildlife watchers, including tourists who come to Texas just to enjoy the natural beauty that our wetlands provide.

A Vanishing Treasure

Texas has lost more than half the wetlands it had before settlement by Europeans. About 7 million acres of wetlands are gone. Many were destroyed by being drained and filled with dirt to use for farming or as land on which

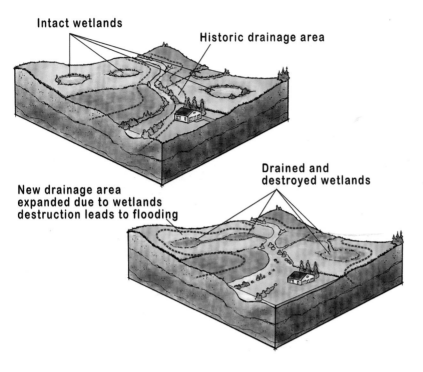

Figure 10.11. Flood control is among the most important benefits provided by wetlands. This benefit often goes unnoticed, because wetlands that store floodwater are often located far from where flood damage occurs. When a wetland is drained and destroyed, the runoff that was once stored in the wetland when it rains has to go somewhere. Rather than being stored and slowly released from the intact wetland, the water flows rapidly into rivers and streams. As more and more wetlands are drained throughout the watershed, with nothing to stop the rush of water entering rivers and streams, the amount of water in them becomes more than can fit. This results in water overflowing the river's banks and flooding adjacent areas. Through conservation of wetlands, properties and people living downstream are protected from flooding. Important aquatic habitat for plants and animals is also protected. Illustration courtesy of Ducks Unlimited Canada.

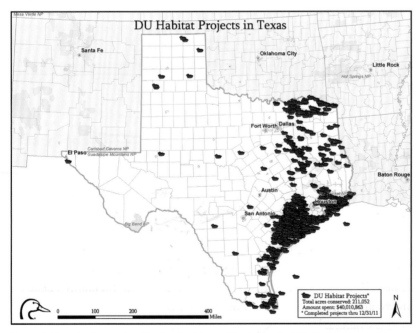

Figure 10.12. Waterfowl and wetlands scientists working for Ducks Unlimited, a conservation organization with members who are also duck hunters, consider the Texas coast to be one of the top-five priority areas for wetlands conservation in all of North America. Many conservationists are working to restore wetlands in Texas. Partnerships for conservation of wetlands between Ducks Unlimited, Texas Parks and Wildlife, the US Fish and Wildlife Service, and others have resulted in over 200,000 acres of wetlands conserved in Texas. Map courtesy of Ducks Unlimited, Inc.

WETLANDS

113

to build our homes and businesses. Many of the wetlands that are left have been partially filled by sedimentation, polluted, or altered to the point they no longer function naturally. State and federal conservation agencies work together with wildlife conservation organizations to protect and restore wetlands (fig. 10.12).

You can help, too, by supporting efforts to protect wetlands no matter where you live. Do your part to keep pollution from entering Texas' waters. You can learn more by visiting a wetland. You can also help watch water quality in wetlands by volunteering as a water quality and habitat monitor with the Texas Stream Team. Unlike many aquatic habitats, wetlands are easy to see and explore. Many plants and animals can usually be seen near the water's edge. Canoeing, kayaking, fishing, and bird-watching are just a few of the fun things you can do while enjoying the beauty of a wetland.

AQUATIC SCIENCE CAREER

Wildlife Technician and Biologist
Wildlife technicians and biologists work with all forms of wildlife and in wildlife habitats, varying from deserts to forests to wetlands. Technicians do a lot of fieldwork, such as capturing and counting wildlife species. Biologists work in the field but also may do laboratory work, such as testing wildlife for fat content or contaminants to see how healthy the animals are. Biologists also do computer-aided studies using information collected from the field. For species such as this mottled duck in the photograph, biologists study how many individuals there are in the population and how well they are reproducing. Technicians usually have at least a bachelor's degree in wildlife or natural resources management. Biologists usually have a master's or doctoral degree in wildlife biology or a related field.

Photograph courtesy of Texas Parks and Wildlife Department.

chapter 11

Bays & Estuaries

Questions to Consider

- How do bays differ from estuaries? How are they similar?
- Why is freshwater inflow important in bays and estuaries?
- What is a hypersaline bay?
- What causes tides? Why is there a high and a low tide? How does this changing flow of water affect aquatic life in bays and estuaries?
- What kinds of plants grow in coastal wetlands?
- Compare the adaptations of the spotted seatrout and the red drum. How do these adaptions affect their life in bays and estuaries?
- Why is the Texas coast important to a bird that nests in Canada or Venezuela?
- What are some of the economic impacts of bays and estuaries?

CHALLENGE QUESTIONS

How can you help maintain healthy coastal ecosystems in Texas? Has your life been affected by bays and estuaries?

Estuaries and bays form where rivers meet the ocean. An **estuary** is a partly enclosed body of water along the coast where one or more streams or rivers enter and mix freshwater with seawater. A **bay** is a body of water partially enclosed by land that is directly open, or connected, to the ocean. In Texas, many bays are also estuaries. There is often no clear point at which the estuary ends and the bay begins. Some estuaries simply extend out into the Gulf where there are no bays. These places are often called **deltas**. The largest delta in the Gulf is formed by the Mississippi River in Louisiana.

Down by the Boardwalk

If you traveled along the almost 400 miles of Texas coastline, you would find more than 2.6 million acres of estuary habitat and 7 major bays (fig. 11.1). Each bay is very different from the next. These differences give rise to the high biodiversity in aquatic ecosystems along the Texas coast.

The coast from Sabine Lake, on the border between Texas and Louisiana, to Galveston Bay is characterized by extensive wide wetlands that go from freshwater, to brackish, to saltwater as you move closer to the Gulf. The coast from Galveston Bay to Corpus Christi Bay consists of large bays and estuaries supplied with freshwater inflow by rivers, but as you move south down the Texas coastline, there is less and less freshwater inflow.

The coast from Corpus Christi Bay to the border with Mexico consists of the upper and lower Laguna Madre. The water in this bay, or **lagoon**, is often **hypersaline**. Salinities there frequently exceed that of average seawater! Why? Laguna Madre has few inlets to the Gulf, little freshwater inflow (no rivers flowing in and little rainfall), and high evaporation. All these factors increase salinity above that of normal seawater. The Laguna Madre is the only hypersaline bay in the United States, and one of only six in the world. Poor conservation has now meant less freshwater flowing into Nueces Bay, causing periods of hypersalinity in that bay as well. Human-caused factors can disturb natural salinity levels, creating the unusual situation of higher salinity where the river enters the estuary than where the bay empties into the Gulf. Most estuaries are fresher near the river and saltier near the ocean.

Figure 11.1. Major bays along the Texas coast. Photographs show a variety of landscapes. Photographs and map courtesy of Texas Parks and Wildlife Department; modified by Rudolph Rosen.

The coast from Galveston Bay all the way to Mexico is characterized by **barrier islands**, which are long, narrow islands of sand that run parallel to the **mainland**. They form the seaward side of many of Texas' bays and lagoons. They shield important wetlands and seagrass habitat along the bay side of the islands (the side facing the mainland) and help protect coastal communities from **storm surges**. The longest barrier island in the world is Padre Island, which is 130 miles long and protects nearly one-third of the Texas Gulf Coast.

The Power of Freshwater Inflows

How much, how, and when freshwater flows into the estuaries and bays in Texas are the biggest influences on life in these aquatic ecosystems. Galveston Bay, Matagorda Bay, San Antonio Bay, and Corpus Christi Bay are like big mixing bowls where freshwater inflows create salinity gradients that expand and contract with seasons, droughts, tides, and floods.

Life in the estuaries has adapted to normal rainfall patterns inland. Estuaries receiving freshwater inflow from East Texas rivers are adapted to higher amounts of freshwater inflow than estuaries in South Texas. Estuaries in Texas need higher freshwater inflows during the late spring and early summer. Along with freshwater, the inflows bring nutrients and sediments that feed fish, wildlife, invertebrates, plankton, and wetland plants. In recent years freshwater inflows have been lower than biologists consider adequate to maintain highly productive aquatic environments. This lower inflow has been due to persistent drought conditions and lack of adequate water conservation in watersheds to maintain enough freshwater flow into the estuaries. The mixing of freshwater with seawater in bays and estuaries varies with the geology, hydrology, and human alterations, such as **dredging** new or deeper channels between the bay and Gulf (fig. 11.2).

Tides also have a large and varying effect. **Tides** are the alternating rise and fall of sea level produced primarily by the combined gravitational attraction of Earth, the moon, and the sun on the oceans. These forces are variable but predictable. The heights and timing of tides vary daily as the relative positions of Earth, the moon, and the sun change. As the level of the waters in the Gulf rises, seawater rushes into the bays from the ocean through passes or inlets. "High tide" occurs when waters reach the highest point, and the waters rushing in are called the **flood tide**. At "low tide," water that came into the bay during the rising tide rushes back out into the Gulf. This is called the **ebb tide**.

Tides in Texas move up and down on average about 1.5 feet, but sometimes they can vary from low to high by as much as 3 feet. On some days tidal fluctuation in water height can be very small, and on other days it can be large. In other parts of the world the tide can change up to 30 feet per day! Tides are different depending on exactly where you are located on the coast. The tides in Galveston Bay at any point in time will be very different from those in Corpus Christi Bay.

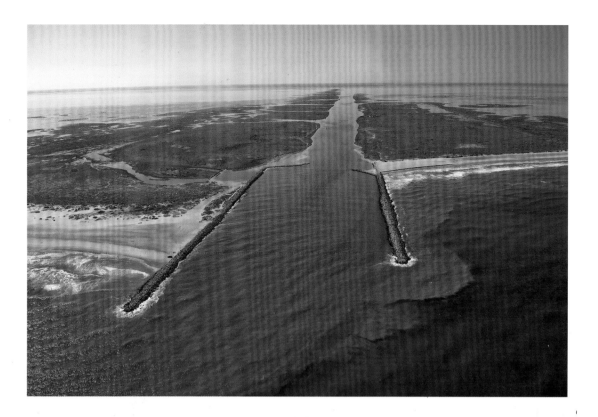

Figure 11.2. The human-made Port Mansfield Channel is an inlet cut through Padre Island 30 miles north of the southern end of the island. This opening between the Laguna Madre (mother lagoon) and the Gulf changed water-circulation patterns in the lagoon, allowing an exchange of waters between the lagoon and Gulf. The new circulation patterns enhanced the productivity of waters in the lagoon, leading to a greater abundance of red drum, shrimp, flounder, and spotted seatrout, as well as increasing the economic value of sport and commercial fishing in Port Mansfield. The channel also allows boats to now travel between the lagoon and Gulf. Photograph courtesy of Texas Parks and Wildlife Department.

Tides and winds help mix the freshwater and saltwater in the estuaries. Add to this the variability of freshwater inflows, the hydrology, and any human-caused alterations that affect flows, and you begin to realize that the mixing of freshwater and saltwater in Texas bays and estuaries is a very complex process. It may be complicated, but it is also one of the most important things to understand when studying the productivity of the aquatic ecosystems in bays and estuaries. Shrimp, crabs, oysters, red drum, and spotted seatrout are some of the most familiar species that have adapted to take advantage of these dynamic ecosystems (fig. 11.3).

Life in Bays and Estuaries

Estuaries and the wetlands in them are where the young of our most important Gulf species live before becoming adults. These areas provide places for young fish, shrimp, and small crabs to hide from predators and find food. Sediment from freshwater inflows settles to the bottom in the quiet waters of the wetlands, where it provides habitat for plants and burrowing invertebrates. The sediments carry tiny bits of detritus and nutrients, such as nitrogen and phosphorus, that feed plants, plankton, and other life. The microscopic plankton are filtered from the water and eaten by oysters that build reefs. The reefs provide more shelter for fish and crabs. Without enough freshwater, sediments, and nutrients, the estuaries could not function as nursery grounds for the fish and shellfish we like to catch and eat. Marine biologists have deter-

Figure 11.3. Spotted seatrout, crabs, oysters, and red drum are among the most important and familiar species for people to catch and eat in Texas bays and estuaries. Photograph by Captain Mike McBride (*top left*); photographs courtesy of Texas Parks and Wildlife Department (*top right, bottom left, bottom right*); photograph by Rudolph Rosen (*background*).

mined that 90% to 95% of all commercially and recreationally important species are found in our estuaries at some stage of their life cycle.

Nature's Nurseries

Estuarine ecosystems contain essential nursery habitats for our seafood. As salinity increases from the river side of the estuary out to the saltier part, **seagrasses** replace the freshwater grasses as cover and places to feed. Seagrass beds are also highly productive nursery areas. The plants grow in large areas or clumps, called "beds," and provide food for microscopic plankton at the base of the food web. The plankton then become food for newly hatched shrimp, fish, and crabs. As these animals grow, the seagrass provides a place to hide from predators. Seagrass beds are therefore great places to look for spotted seatrout, red drum, and other interesting species. Seagrass beds reduce

SEAGRASS GROWS IN TEXAS ESTUARIES

Preserving the three species of seagrass in Texas maintains critical habitat for saltwater fish, shrimp, and crabs. All three species need clear, unpolluted water, and their presence or absence is an indicator of the health of the bays and estuaries.

Turtlegrass

Turtlegrass occurs in deeper waters and high salinities in Aransas, Redfish, and Corpus Christi Bays and the lower Laguna Madre. It is the dominant seagrass in the most southerly part of the lower Laguna Madre. It is a slow grower and takes time to recover from any disturbance or stress.

Shoalgrass

Shoalgrass grows in high-salinity waters of all bays south from East Matagorda Bay. It is especially abundant in the upper Laguna Madre but is declining in the lower Laguna Madre and in West Galveston Bay.

Widgeongrass

Widgeongrass is abundant in low-salinity waters of all upper bay areas along the Texas coast. It often occurs in spring, growing with shoalgrass, and commonly occurs in brackish ponds near the coast.

Photographs courtesy of Texas Parks and Wildlife Department (*top*); Warren Pulich (*middle*); Wes Tunnell (*bottom*).

erosion because the plants' roots help bind the soil together. They even act as biofilters by settling out sediments.

Unfortunately, many seagrass beds have been lost. More than 90% of the seagrass beds in Galveston Bay have been destroyed because of storms, hurricanes, disease, **toxic algae blooms**, and development. Dredging, boat propellers, and high currents stir up the water and raise sediments that make the water more turbid. This can block sunlight from reaching the seagrasses and prevent photosynthesis, which seagrass needs to survive. As of 2013 about 80% of remaining seagrass habitat in Texas was located in the Laguna Madre. The Laguna Madre shoreline is protected from development by large ranches on the mainland and by Padre Island National Seashore, which protect the barrier island from development. Biologists are working to restore seagrass to some locations in Texas' bays and estuaries (fig. 11.4).

Another key habitat providing essential nursery area for our most important seafood species are oyster reefs. Large numbers of oysters join together to create massive underwater structures called oyster reefs. Oyster reefs provide habitat for tiny **periphyton** and zooplankton, **macroinvertebrates** such as crabs, small fish, and larger fish looking for crabs and smaller fish to eat. An oyster reef creates an entire aquatic community.

But oysters are more than just shells and habitat. They are also one of our most important seafoods. Millions of pounds of oysters are harvested each year to be served in restaurants and sold in grocery stores. Many people are employed in harvesting these oysters, and it is an important part of the coastal economy.

To many people, oysters are just an item on the menu at a restaurant, but what they may not know is that oysters help clean up pollution. Oysters are filter feeders that use their gills to strain tiny food particles such as plankton, as well as detritus and sediments, out of the water. This helps clean the water in our estuaries and bays. Sometimes they ingest pollutants as they filter the water. Although this helps make the environment healthier for other aquatic life, including other important seafood species, it may make the oysters unsafe for people to eat. The Texas Department of State Health Services determines which areas are too polluted for oyster harvesting, because oysters there have become contaminated and unsafe to eat.

Adaptations of Red Drum and Spotted Seatrout to Their Habitats

Two of the most popular saltwater fish to catch and eat in Texas are red drum and spotted seatrout. Healthy estuaries allow both species to thrive. Yet like shrimp and many other species that live in the estuaries, red drum and spotted seatrout lead a dual life. They also spend part of their time in the deeper, open waters of the Gulf. The way they split their lives between the two ecosystems shows how two closely related species can have very different adaptations to survive in the variable conditions of Texas' bays and estuaries.

Figure 11.4. Volunteers help Texas Parks and Wildlife biologists replant and restore seagrass to Texas bays and estuaries. Photographs courtesy of *Corpus Christi Caller-Times* (*top*) and Texas Parks and Wildlife (*bottom*).

SPOTTED SEATROUT. Spotted seatrout **spawn** in coastal bays, estuaries, and lagoons. They prefer to do this in shallow, grassy areas where eggs and larvae have cover from predators. The eggs hatch in about 18 hours, and the larvae feed mostly on zooplankton. In about 20 days they become juveniles, or miniature versions of the adults. As they grow, they begin feeding on benthic invertebrates and tiny shrimp, then shrimp and small fish, and when they get large, they feed almost exclusively on other fish. Throughout their life spotted seatrout prefer to remain near seagrass beds and oyster reefs as they look for prey. But if the water gets too cold during fall and winter, seatrout move into the Gulf of Mexico. The open ocean is such a huge amount of water that it does not change temperature as fast as the shallow estuaries do. As water temperatures warm in the spring and summer, the fish return to the shallows of the bays and estuaries. This species has evolved a strategy of using the food-rich cover of seagrass beds and oyster reefs as long as the shallow waters are warm enough.

RED DRUM. Adult red drum live in the waters of the Gulf of Mexico. They spawn in high-salinity waters near an entrance to a bay where high tidal current flow will carry the eggs and larval fish into the bay and estuary. Young red drum in the estuaries are most abundant in water one to four feet deep where there are seagrass beds. They are also found around other structures, such as along jetties and pier pilings. They are aggressive, opportunistic feeders and grow fast. Young fish feed on small crabs, shrimp, and marine worms. As they grow larger, they feed on larger crabs, shrimp, and smaller fish. Red drum stay in bays and estuaries until they become mature. This may occur as early as three years of age or as late as six years. When mature, they move into the Gulf of Mexico and live there the rest of their lives, usually staying within about five miles of shore.

Destination: Texas

Coastal barrier islands and wetlands provide habitat and protection from storms for millions of migrating waterfowl, shorebirds, and neotropical migratory birds from throughout the Western Hemisphere. Birds migrating in the spring often travel thousands of miles. Some even fly all the way across the Gulf of Mexico before landing on Texas shores.

The lives of these winged visitors are just as linked to the health of our bays and estuaries as are the lives of the fish and invertebrates that cannot just fly away. Over 75% of the world's population of redhead duck winters in the Laguna Madre (including the portion in Mexico). They feed in the shoalgrass beds and coastal ponds. The winter home of the endangered whooping crane, one of the rarest birds in the world, is located in San Antonio Bay. Whooping cranes feed almost exclusively on blue crabs. But crabs in the bay are dependent on the amount of freshwater inflow from the Guadalupe and San Antonio Rivers. Low inflows already have resulted in loss of cranes. Texas also has the largest colonies of roseate spoonbills and reddish egrets in the world, and

Photograph courtesy of Texas Parks and Wildlife Department.

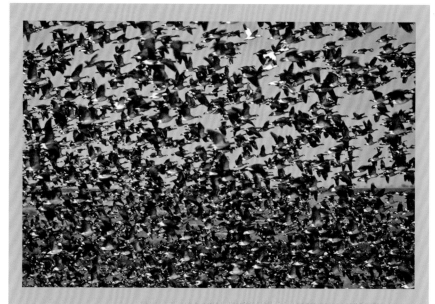

SEE TEXAS' GREAT WINGED MIGRATIONS

The estuaries, wetlands, and coastal islands in Texas sit at the crossroads of one of the greatest annual migrations of wildlife anywhere in the world. Africa has its great migrations of hooved animals that cross the continent's vast plains. Here in America we have great winged migrations. These consist of millions of waterfowl, wading and shorebirds, and neotropical migratory birds.

Millions of neotropical migratory birds make first landfall on the Texas coast after long-distance flights across the Gulf between North and South America. In foul weather the birds may arrive totally exhausted, looking for a place to land, rest, and feed to regain energy for continued flight.

Texas coastal wetlands also serve as the key wintering ground for millions of North American waterfowl. In early spring, these birds fly thousands of miles north of Texas to lay their eggs and raise ducklings in the summer. They go to northern states, Canada, and even the Arctic. When it starts getting cold in early fall, the young are ready to fly. The adults along with the young birds return south to Texas, where they feed all winter long. Coastal Texas is one of the most important staging, wintering, and breeding areas for waterfowl and other migratory birds on Earth.

Each year, thousands of bird-watchers come to see these great migrations. Almost 500 bird species have been seen in the lower coast alone. You can see the birds, too. Visit a Texas wildlife management area or national wildlife refuge, the World Birding Center, or any one of the many local parks and nature centers on the coast.

our barrier islands are nesting areas for the critically endangered Kemp's ridley sea turtle.

The Importance of Bays and Estuaries

More than one-third of the human population and about 70% of the industry and jobs in Texas are located within 100 miles of the coast. Almost 5 million people live in the 18 counties adjacent to the Gulf. More than one-third of all the oil and gas production in the entire United States takes place on the Texas coast (fig. 11.5). Texas also has more marine commerce than any other state, which involves ships carrying various goods. Texas has 10 large seaports and includes more than 420 miles of the Gulf Intracoastal Waterway. This waterway is like a giant highway for ships (fig. 11.6).

Texas estuaries annually produce more than 100 million pounds of seafood valued at $150 million to $250 million per year. Saltwater recreational fishing generates an estimated $2 billion, and coastal tourism provides $5.4 billion in economic activity annually (fig. 11.7). Activities tied to the health of our bays and estuaries have contributed to making tourism the third largest industry in Texas, after oil and gas production and agriculture. Nature lovers from all over the world visit the Texas coast just to see rare species. Tourism for just the whooping crane alone results in over $6 million to Texas' coastal economy. These values are easy to add up, but there are less apparent benefits, such as how healthy estuaries provide **ecosystem services** we all depend upon. These include treating municipal wastes, diluting industrial discharges, and supporting wetlands that protect our communities from storm surges.

Over the last 100 years our coastal ecosystems have changed dramatically. About 50% of Texas' coastal wetlands have been filled in, eroded away, or otherwise destroyed. Up to 60% of the seagrass beds are gone, and over 50% of the oyster reefs no longer exist. These habitats depend on an adequate supply of freshwater coming out of the rivers each year, but in some years more water than even flows in the rivers has been promised to users, such as cities and farmers, upstream from the coast. While there are efforts by conservation groups and government environmental agencies to restore some coastal habitats (see figs. 10.12 and 11.4), loss of bay and estuary habitat for aquatic

Figure 11.5. Oil and gas development has long been a mainstay of the Texas Gulf Coast economy, providing jobs and billions of dollars in economic activity by delivering our nation needed energy supplies. This development has impacted the coast in many ways. Many of our industrial areas, business centers, and communities have been built on what were once coastal wetlands, estuaries, and islands. Bay and estuary bottoms are crisscrossed with channels to allow for boat commerce and promote petroleum development, but this drastically alters bay and estuary water flow patterns. We depend upon bay and estuary waters to treat waste and assimilate pollution. We are fortunate that Texas bays and estuaries are so resilient. Photograph by Jace Tunnell.

Figure 11.6. The Gulf Intracoastal Waterway is a protected channel used primarily by large barges and tugboats to transport goods all along the Texas coast and eastward to Louisiana and beyond. The channel is continually dredged, or kept deep, by digging out sediments that fill the channel. This allows larger vessels to use it without scraping bottom. Larger and smaller boats, such as oil field and fishing boats, are protected from large waves, storms, and strong currents that may form in the Gulf. Photograph courtesy of Texas Parks and Wildlife Department.

Figure 11.7. Texas bays and estuaries are often windy places due to thermal breezes. These are winds created as a result of temperature differences between the land and ocean. In spring months, the land heats up quickly during the day (creating lower air pressure over the land), making it much warmer than the ocean waters, which take much longer to warm up. This creates thermal winds during many afternoons along the Texas coast as cooler (higher air pressure) ocean air flows toward the warmer land, creating an onshore wind. These winds can reverse at night, if the land cools off and the air on land becomes cooler than the air over the ocean. Windy days make Texas bays and estuaries places many people come to enjoy water sports that require wind, such as sailing, kiteboarding, and windsurfing. Photograph courtesy of *Corpus Christi Caller-Times*.

species is the greatest threat to our coastal economies and marine life. Compounding the challenge is that our coastal region is one of the fastest developing in the world and is vital to the economy of the United States.

CLASS PROJECT: RECYCLE AN OYSTER REEF

Most people know oysters have a hard outer shell and live on the bottoms of bays and estuaries in Texas. But what many people do not know is that oysters do not start out in a shell. After hatching from tiny eggs, the soft oysters drift in the water, pulled along by currents in the bay for about three weeks. They then find a hard surface upon which to attach. Once attached, they begin growing their shells.

Oysters in Texas mostly grow on the shells of other oysters, forming an oyster reef over time. When oysters are harvested for food, the whole animal, shell and all, is removed. The soft oysters are removed from the shells when prepared as food in restaurants and by seafood processors. Where do all the shells go?

A great class project is to help restore these oyster shells to oyster reefs so new oysters will have a place to attach and grow. This is called oyster shell recycling. In cooperation with Texas Parks and Wildlife biologists, oyster shell is returned to oyster reefs to keep the reef healthy and supply oysters for the future. Without recycling, the shells will be thrown away and the oyster reef will not be able to regenerate new oysters.

In the photographs, high school students volunteer in a "Sink Your Shucks" oyster shell recycling program sponsored by the Harte Research Institute at Texas A&M University–Corpus Christi. Restaurants supply the shucks.

Photograph courtesy of *Corpus Christi Caller-Times* (*left; top and bottom*) and Harte Research Institute for Gulf of Mexico Studies (*right; top and bottom*).

AQUATIC SCIENCE CAREER

Benthic Ecologist

There are thousands of invertebrate species in aquatic systems, from mountaintop headwaters to the greatest depths of the oceans. Benthic ecologists study the invertebrates and the communities of invertebrates that live on the bottoms of streams, lakes, rivers, estuaries, bays, and the ocean. They do research experiments, take counts and measurements, look at the effects of pollution, and determine the health of invertebrate communities such as are found on oyster reefs. Benthic ecologists usually have a master's or doctorate degree in biology or ecology.

Photograph courtesy of Texas Parks and Wildlife Department.

chapter 12

∎ ∎ ∎ ∎ ∎ ∎

Oceans: The Gulf of Mexico

Questions to Consider

- Which states share Gulf of Mexico waters? Which other countries share the Gulf?
- What are some of the industries in the Gulf? How can people in these industries help keep Gulf waters healthy for aquatic life?
- What influence does the Mississippi River have on the Gulf?
- What is a hypoxic zone and its impacts? How are hypoxic zones formed? How can they be prevented?
- What might you see on a Texas beach?
- What are some of the ecosystems in the Gulf, and what kinds of organisms would you find in them?
- How are oil and gas platforms similar to coral reefs?
- What are currents? What do they do in the Gulf?
- How can you help marine mammals or sea turtles?

CHALLENGE QUESTIONS

What do you think scientist Sylvia Earle meant when she referred to the Gulf of Mexico as "America's Sea"? What does it mean to you?

The Gulf of Mexico is the ninth-largest **ocean basin** in the world, covering almost 600,000 square miles. This is twice the size of Texas. At its deepest point the bottom is 2.7 miles underwater, but most of the Gulf is much shallower. About 60% of the Gulf is less than 700 feet deep.

The Gulf is a very accessible body of water. It is surrounded to the north, east, and west by five US states (Florida, Alabama, Mississippi, Louisiana, Texas), to the west and south by six Mexican states (Tamaulipas, Veracruz,

Tabasco, Campeche, Yucatán, Quintana Roo), and to the southeast by the island of Cuba. Watersheds covering more than 60% of the United States drain into the Gulf, including 33 major river basins and 207 estuaries (see fig. 6.3). Watersheds covering about half of Mexico and one-third of Cuba also drain into it. Because the Gulf is surrounded by North and Central America, it has been called "America's Sea."

Rivers carry the runoff and other waters from Texas' watersheds to our estuaries, so estuaries tell us a lot about how well we are protecting our water resources. Every day, you and I affect the health of the Gulf ecosystem. Cooking, cleaning, watering the lawn, flushing, and even driving the car can cause water pollution if done without regard to the impact on the environment. When multiplied by activities that millions of us do who live in the river basins that empty into the Gulf, it is easy to understand that how we treat our watersheds affects the health of Gulf ecosystems.

The Gulf is one of the most productive waters in the world for aquatic organisms. It provides habitat for many of our most important seafoods. The Gulf also holds one of the world's largest reserves of oil and gas, leading to an extensive petroleum industry. The Gulf's combination of ecosystem productivity, industrial importance, and accessibility has made it one of the most important waters to people on Earth as well as one of the most threatened by human actions and neglect.

A Big Bowl Full of Saltwater

The Gulf is shaped like a giant wide-brimmed bowl and filled with saltwater. It is shallow around the edges but deep in the middle. The edge of the Gulf, starting from the coastline and moving toward deeper water, is full of wetlands, estuaries, and bays, opening out to wide and shallow shelves that gradually slope into deep waters. The floor of the Gulf is mostly a vast expanse of soft mud. The freshwaters that flow into the Gulf greatly affect the health and productivity of the aquatic life there.

The Ultimate Downstream Aquatic Habitat

One of the most important influences on aquatic ecosystems is the Mississippi River, which accounts for nearly 90% of all freshwater inflow to the Gulf. The watershed for the Mississippi River is more than six times larger than Texas. It includes 31 states and two Canadian provinces (fig. 12.1). Like an enormous water hose, the Mississippi River pours freshwater into the Gulf.

The natural flow from the Mississippi River has always influenced productivity in the Gulf. Freshwater inflow provides nutrients that are carried by currents throughout the Gulf. The nutrients promote growth of phytoplankton. These are primary producers that form the base of an extensive food chain that includes zooplankton, macroinvertebrates, fish, whales, sea turtles, seabirds, and many other forms of marine life.

Figure 12.1. The Mississippi River watershed is the dominant source of freshwater and nutrients to the northern Gulf of Mexico. Spring rains and snowmelt in the Mississippi River watershed start the chain of events that results in formation of a hypoxic zone, where water near the bottom of the Gulf contains very low amounts of dissolved oxygen. The zone expands from the mouth of the river and into the Gulf along the coasts of Louisiana and Texas each summer. Hypoxic zones occur elsewhere in the world's oceans, but the one in the Gulf is now the second largest on Earth, sometimes extending all the way from Texas to Florida. Illustration by Rudolph Rosen.

Human activities have added wastes, pollutants, fertilizers, and extra sediments to the flow of the Mississippi River. As long as the Gulf remains healthy, it can absorb the impact of some level of nutrients and pollution. Today, as much as 159 million tons of sediment may flow into the Gulf each year. Increased amounts of nitrogen and phosphorus fertilizers from agriculture in the Mississippi River's watershed have caused overenrichment and direct pollution of Gulf waters.

Extra nutrients have upset the natural balance of aquatic production in Gulf waters adjacent to Louisiana and Texas. The nutrients create a rapid, massive growth of phytoplankton at the water's surface in the summer. This results in a biomass of primary producers far beyond what would occur naturally, often called an **algae bloom**. The increase in phytoplankton then affects the Gulf food chain, increasing food for zooplankton and other aquatic life. But the amount of phytoplankton produced in such a short time is well beyond the capacity of primary consumers to graze it down to a balanced level. Phytoplankton have a relatively short life span, so much of the phytoplankton dies before it can be consumed. When the phytoplankton organisms die, they sink down to the ocean floor where decomposers, such as bacteria, break them down.

At the time of year this usually happens, the water column is stratified, meaning that temperature and salinity differences between surface and bottom water layers prevent the layers from mixing. This isolates bottom waters from being resupplied with oxygen from the surface. The plankton that has sunk to the bottom is decomposed by bacteria, but large amounts of dissolved oxygen are consumed by the bacteria, and the dissolved oxygen is quickly depleted. The result is creation of a **hypoxic zone**, often called the dead zone. This is an area of very low to no dissolved oxygen. Organisms such as fish and shrimp that are capable of swimming away may leave the area, but life that lives in or on the bottom has nowhere else to go. Many species can experience stress or die as a result of this low dissolved oxygen. **Hypoxia** adversely affects production of seafood and other aquatic life because food webs are disrupted and organisms at all trophic levels are harmed. Hypoxia can last for several months until the water layers mix, which can happen when storms occur or when the surface water cools in fall and winter.

Life around the Edges: The Seashore

The seashore is the first thing many of us see when we visit the Gulf. We may walk along the sandy beach, swim or fish, or just watch as the waves come rolling in (fig. 12.2). Sandy beaches make up more than 98% of the Texas shoreline. The beaches vary from being part of the mainland, to creating peninsulas, to forming barrier islands. Texas has the most extensive barrier islands in the Gulf, including Padre Island, the longest barrier island in the world. It is 130 miles long and protects nearly a third of the Texas coast.

Beaches in the western Gulf of Mexico, where Texas is located, are considered to be moderate- to high-energy beaches where there are large waves most of the year. The eastern Gulf is a low-energy area, which has smaller waves, on average. This makes the Gulf Coast of Texas a better place for surfing and wind sports, such as windsurfing and kiteboarding, than the Gulf Coast of Florida. Just off the beach, especially in the moderate- to high-wave-energy areas, are one or more sandbars and troughs that get progressively deeper as you go offshore. This explains why, as you wade into the Gulf from a Texas beach, the water quickly gets deeper, but if you keep going, it gets shallow again, and then becomes deeper again as you wade out farther.

Aquatic life is generally abundant where the waves crash into the shore, but this life is not easily seen. Most species here live by burrowing into the shifting sands of the bottom. Beach clams, marine worms, and sand dollars are common in the surf zone. In this area, the swash zone, the waves wash back and forth along the beach. The beach slopes steeply up to a flat area that generally stays dry. This is the backshore, where people usually walk and lie on beach towels. This is also the part of the beach where you can find sand crabs and occasionally nesting sea turtles.

Sand dunes make up the zone behind the backshore. They vary in height from only a few feet up to the largest in the Gulf at 30 feet on Padre Island. They are usually covered with grasses that help hold the sand in place. The dunes help protect areas behind the beach from wind and storms. Sandy beaches are important to the Texas economy because many people enjoy

Figure 12.2. Beaches and educational boat tours offer great opportunities for family fun and education about the Gulf of Mexico ecosystem. Photographs courtesy of Texas Parks and Wildlife Department (*top*) and Corpus Christi Caller-Times (*bottom*).

going to the beach and the many beach parks along the Texas coast. All of these are great places to visit, go fishing, or go surfing. In addition, these parks help protect wildlife and aquatic life.

Life in the Gulf

Life in the Gulf is diverse because there are many different ecosystems and habitats that support many kinds of organisms and species. Even in the deepest and darkest reaches of the Gulf there is life. **Chemosynthetic organisms** form deep in the Gulf near **cold seeps** where hydrogen sulfide, methane, and other hydrocarbon-rich fluids seep from the Gulf's floor. Organisms living there use these resources in chemical reactions that produce energy. Other communities deep in the Gulf waters consist mostly of bacteria and

WHAT MAKES WAVES?

Waves are usually formed when water is pushed by winds. In other words, wind energy creates waves. Waves can also be formed by energy from earth movements, such as an underwater earthquake. The water itself does not really flow along with the wave but instead moves in a rolling fashion as the energy passes through it. This is similar to your sending a "wave" down a rope held between two people. The rope just moves in place; it does not actually travel in the direction of the wave.

A wave can travel for thousands of miles, but when it nears shallow water, such as shore, the bottom part of the rolling water begins to touch the ocean bottom. The lower part of the wave slows a bit while the upper part keeps going. At this point, the faster water on top begins to tip over, giving rise to the curl on large waves as they come to shore and "crash" on the beach.

Illustration by Rudolph Rosen (*top*); photograph courtesy of Texas Parks and Wildlife Department (*bottom*).

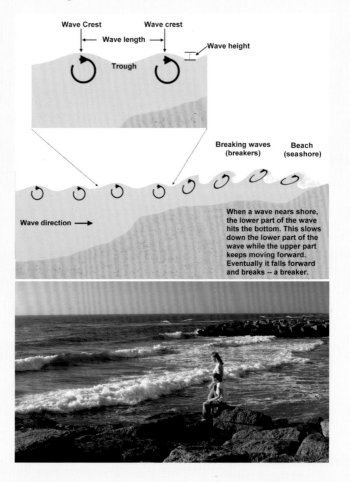

CHAPTER 12

tiny benthic organisms. Macroinvertebrates abound, such as crabs, mollusks, shrimp, and jellyfish.

Currents move throughout the Gulf, forming big loops. These currents help create areas where water from the bottom rises to the surface. This upwelling of water causes nutrients to mix in the water. The increase in nutrients in surface waters enables primary producers, such as phytoplankton, to grow and become food for zooplankton and filter feeders, such as the giant whale shark or Bryde's whale. The plankton allows for growth of smaller filter feeders as well, such as menhaden, which travel the Gulf in large schools. These become prey for large **pelagic fish** that live near the surface, such as tuna and mackerel.

Sometimes toxic or harmful algae blooms occur when the species of plankton involved in algae blooms have toxic characteristics. These are sometimes referred to as red tides because of a red or reddish-brown color caused by some phytoplankton species. The blooms can kill fish and marine mammals. They can also cause health problems for people exposed to the water or to the blooms when they reach close to shore.

On the coral reefs there is an abundance of invertebrate species that use the hard surface as a place of attachment. This life attracts small fish, which attract larger predators, such as groupers, amberjack, and snappers. The beach zone, oil and gas platforms, and deep canyons are just some of the areas that add to the diversity of habitats for Gulf life. Species normally associated with land are also found in the Gulf. There are turtles and birds with special adaptations for a life at sea, such as the sea turtles' flippers and the seabirds' "tear" glands that allow them to excrete excess salts.

Coral Reefs

Of the Gulf's many ecosystems, coral reefs stand out as the most biologically diverse, productive, and complex. They are economically important as sources of food and medicinal products, and they provide shoreline protection in some places. They also are a source of great natural beauty and provide significant tourism opportunity. However, degradation and loss of this ecologically and economically valuable habitat are of worldwide concern. Coral reefs are among the most endangered marine ecosystems on Earth.

Coral reefs are tropical, shallow-water ecosystems. There are only two true coral reefs in the northern Gulf of Mexico. These lie about 100 miles off Texas within Flower Garden Banks National Marine Sanctuary. Flower Garden Banks received its name from the colorful collection of reefs that resemble a flower garden when viewed from the surface. They sit atop salt domes, which rise up from the seafloor. The reefs are about 60 to 400 feet below the surface. Texas' coral reefs are the northernmost coral reefs in the Gulf of Mexico. Fortunately, they are some of the healthiest coral reefs in the Western Hemisphere.

Flower Garden Banks is visited by a wide array of aquatic life, including numerous species of rays and sharks, sea turtles, and marine mammals. More than 23 kinds of coral and approximately 400 species of reef invertebrates inhabit the banks. This includes at least 27 species of sponges, 20 species of **polychaetes** (marine worms), 162 species of mollusks, and 36 species of **echinoderms** (such as sand dollars). More than 280 fish species thrive in the reefs, including snapper, hogfish, groupers, puffers, and angelfish. Whale sharks visit in the summer, and in winter you can see schools of large hammerhead sharks and rays. Visitors regularly see a resident population of more than 70 manta rays at the sanctuary.

"Artificial" Reef Ecosystems

Vast deposits of oil and gas rest under the Gulf's waters. Extracting petroleum products is the largest economic industry in the Gulf today, supplying 25% of gas and 10% of the oil produced in the United States. To obtain these natural resources, there are about 4,000 oil and gas platforms in coastal and offshore Gulf waters. Although trash or pollution from oil and gas drilling rigs and production platforms has sometimes caused pollution, the platforms add to Gulf productivity by creating small but ecologically diverse habitats where none would exist otherwise (fig. 12.3).

The Gulf of Mexico contains thousands of species of animals, algae, and other life that need hard surfaces to cling to in order to complete their life cycles. Since the Gulf has very little naturally occurring reef or hard structure habitat, human-made structures such as oil and gas platforms give invertebrates such as corals, sponges, polychaetes, mollusks, echinoderms, and other animals the hard surface they need to grow and thrive. Energy from the organisms that accumulate over time then flows up the food chain, as primary producers feed consumers, and predators feed on prey. New habitat is then created for larger predator species, such as snapper, grouper, mackerel, shark, and other fish. Even sea turtles benefit from the new feeding opportunities. This habitat, in turn, provides fishing and diving opportunities for people.

You may wonder what happens to these platforms when they get old. Many continue to enhance the Gulf's productivity. At the end of a platform's functional life it may be cleaned and toppled into the water in place. Other platforms may be cut into large pieces and transported to another spot in the Gulf where they are sunk to the bottom, usually along with other old platforms, to build an underwater artificial reef. There are now more than 150 production platforms making up these reefs.

These new reefs provide habitat to many native species, as well as invasive species. Lionfish and orange cup coral, which are invasive species, have moved into new areas in the Gulf because of the new habitats formed by the artificial reefs. The orange cup coral is now the most commonly found coral species on many of the platforms.

Figure 12.3. Marine life on coral and artificial reefs in the Gulf is diverse and beautiful. Anglers enjoy fishing on the artificial reefs, and underwater divers visit Texas coral reefs at Flower Garden Banks National Marine Sanctuary. Recreational fishing and diving charters are available to visit these areas.

In the photo on the left, an underwater photographer takes a picture of sponges and corals attached to a leg of an oil and gas platform. On the right from top to bottom are pictured a reef fish, polychaete, echinoderm, and a stony coral, all found on the platform's hard underwater structures. Photographs courtesy of Texas Parks and Wildlife Department; modified by Rudolph Rosen.

Marine Mammals

The Gulf of Mexico is home to 28 species of marine mammals. The West Indian manatee, or sea cow, is an herbivore that inhabits coastal rivers, estuaries, bays, and nearshore coastal waters. This species is rarely seen in Texas but does frequent all of the other Gulf states. All other marine mammals in the Gulf are **cetaceans**, or whales and dolphins. The most common is the Atlantic bottlenose dolphin, which are often seen in bays and estuaries in Texas. Other species include the Atlantic spotted dolphin, beaked whales, spinner dolphins, and even killer whales. The biggest marine mammal in the Gulf is the sperm whale. It occurs primarily in mid-depth waters off Louisiana and Texas.

For reasons not well understood, whales and dolphins occasionally get stranded on land. Stranded animals are sometimes found to be ill or injured, but many times the animals appear perfectly healthy. Volunteers often try to help rescue stranded marine mammals. Increased education of the public, active management, and conservation efforts have helped ensure healthy populations of these species in the Gulf.

Sea Turtles

There are seven species of sea turtle in the world, five of which live in the Gulf. All have nested on Texas beaches, some perhaps in large numbers historically, and all are now listed as threatened or endangered. Harvesting of the eggs for use by people, killing adult turtles for meat and household products, and incidental capture in shrimp trawl nets are the biggest reasons sea turtle populations have dropped so low. Today, no harvest of eggs or adult turtles is allowed, and special turtle excluder devices are required on commercial fishing nets to help prevent harm to turtles. Recovery of sea turtles will take many years.

Efforts to protect nesting sea turtles on the sandy beaches of Texas help make sure turtles are successful in laying eggs and hatching baby turtles. The inshore and nearshore Gulf of Mexico waters in Texas now provide important habitat for three sea turtles: Kemp's ridley, green, and loggerhead. Special help is going to the Kemp's ridley, which is the most endangered of all sea turtles. This is the smallest sea turtle in the world, weighing about 100 pounds when fully grown. It tends to stay in shallow water less than 165 feet deep and prefers to eat crabs but will also eat shrimp, clams, jellyfish, and **tunicates**.

Like all other sea turtles, the male Kemp's ridley after hatching spends its entire life in the water, while the female comes ashore only to nest. For the Kemp's ridley, this takes place between April and mid-July. The female nests by crawling up the sandy beach to a point beyond the reach of the waves, where she digs a large hole in the sand with her back flippers. There she lays about 100 eggs, covers the eggs with sand, and immediately returns to the

Figure 12.4. After hatching, young Kemp's ridley sea turtles enter the Gulf, where they are swept from where they hatch by currents, sometimes as far north as Nova Scotia in the Atlantic Ocean. When they mature, they return to the Gulf. Adult Kemp's ridleys are found almost exclusively in the Gulf of Mexico, and most mature females dig nests and lay eggs on the same beaches where they were hatched. Volunteers help protect Kemp's ridley nests and watch over newly hatched turtles as they move down the beach to the ocean. Photographs by Jennifer Idol, The Underwater Designer.

water and swims away. The whole process takes about 45 minutes. The eggs hatch about 45 to 60 days later. The young crawl out of the sand and quickly move down the beach to escape predators (fig. 12.4). Once they reach the water, they swim through the surf zone and into the Gulf, where they are largely carried away by the currents.

It takes about 10 to 20 years to reach maturity, and most breeding females nest at the same beach where they hatched. Once they start nesting, most return to the same beach each time they are ready to lay their eggs. Because of this instinct to return to the same beach, nesting areas are well known and are now protected.

The primary nesting ground for the remaining population of Kemp's ridley sea turtles is a 16-mile stretch of beach in Tamaulipas, Mexico. Biologists at Padre Island National Seashore are now reestablishing and protecting a nesting beach on Padre Island. They locate nests and place the eggs in an incubation facility or a special beach enclosure to protect them from natural or human-caused threats. When the eggs hatch and the young are ready to leave the nest, they are released on Padre Island and guarded from predators as they enter the surf zone.

Fish and Fisheries

The Gulf is a place of incredible biodiversity with over 1,500 species of fish calling it home. But it is the seafood that comes from the Gulf that many people know best. Commercial fisheries annually catch over 1.5 billion pounds of seafood. Shrimp and oysters are the predominant **shellfish** harvested in Texas (fig. 12.5). Catches in the Gulf account for 70% of all the shrimp and oysters that go into grocery stores and restaurants across the United States. Recreational fishing is also important, including catches of flounder, snapper, drum, seatrout, mackerel, shark, and tarpon. Of all the people who go fishing in saltwater in the United States, 45% fish in the Gulf of Mexico.

An Ecologically and Economically Sustainable Gulf

The Gulf is one of the most productive yet ecologically threatened bodies of water in the world. It is an area of intensive oil and gas development, has some of the world's largest commercial and recreational fisheries, and contains many of the largest ports and most active **shipping lanes** in the United States. The employment provided in these industries and economic value to Texas and the country are enormous. Yet we have lost many of our coastal wetlands, seagrass beds, and oyster reefs. In Texas, reduced freshwater flow into estuaries and bays reduces the amount of certain kinds of habitat needed by many of the Gulf's most important species. The Mississippi River, which provides the greatest amount of freshwater inflow into the Gulf, now carries such high loads of nutrient contaminants that large hypoxic areas develop

in nearshore waters off the Louisiana and Texas coast each year, stressing, or even killing, aquatic life.

Yet the Gulf is amazingly resilient. Sea turtles that just a few years ago were on the brink of extinction are coming back, thanks to protective efforts in the United States and Mexico. Fishing regulations, like banning use of certain types of fishing gear, have allowed several species of fish to increase in number and size. With about 1,000 natural seeps of oil into the Gulf totaling almost 50 million gallons of seepage per year (the size of an oil supertanker), the Gulf is naturally filled with bacteria that can feed on and break down spilled (or seeped) oil. This natural resilience has allowed the Gulf to bounce back sooner than probably any other body of water in the world from the damaging effects of accidental spills. Scientists will continue to study the Gulf for long-term or cumulative impacts.

Increasing knowledge about how the Gulf works is leading to better conservation decisions on land and in the Gulf. We now know that we can have a healthy economy and a healthy environment in the Gulf if we work together for that goal.

Figure 12.5. Brown, pink, and white shrimp are estuary-dependent species, meaning they use the estuary and bay for food, growth, and shelter during a portion of their life. But the adult shrimp live and spawn in the Gulf of Mexico. Female shrimp release 100,000 to 1 million eggs that hatch within 24 hours. The young shrimp develop through several larval stages as they are carried shoreward and into bays and estuaries by winds and currents. The young shrimp grow rapidly in the shallow waters. Once they reach juvenile and subadult stages, at about three inches in length, they move into the Gulf for the rest of their lives. Photographs courtesy of Texas Parks and Wildlife Department.

THE CONNECTION BETWEEN SEAWEED, JELLYFISH, AND BEACH TRASH IN TEXAS

Beachgoers in Texas often remember encounters with seaweed, jellyfish, and trash found on the beach. Believe it or not, all three are frequent features of Gulf Coast beaches for the same reason. All are carried along by currents and winds that push them onto Texas beaches. Massive currents swirl about in the giant basin that is the Gulf. As happens when you stir liquid contents in a big bowl, the water in the Gulf moves in a definite direction. This water movement, or current, carries along with it whatever floats in the water. Currents in the Gulf move toward Texas from both the north and south. The currents combine with winds that blow toward Texas. This helps push animal passengers as well as any floating trash or seaweed onto our beaches.

At times Texas beaches may contain a large amount of sargassum, a brown seaweed. Although it may look and smell yucky, this seaweed actually helps build up the beach by acting to hold sand in place. Jellyfish are another passenger in the currents' continuous journey because they are free-floating animals. While some species of jellyfish can give swimmers an unpleasant sting, trash gives everyone an unpleasant experience.

Jellyfish and seaweed are a natural part of the Gulf ecosystem, but the trash is not. Where does trash come from? It comes from all over the Gulf, from other states, from Mexico, from storm sewers that empty into the Gulf, and from the rivers draining into the Gulf, such as the Mississippi River. It comes from ships and oil and gas platforms far out in Gulf waters. It floats northward to Texas from Mexico and southward from Louisiana. The amount of trash that washes to shore is enormous. Sometimes sea turtles and other species that eat jellyfish mistake clear plastic bags or other trash in the water for food and eat the trash. This can cause injury or death because the plastic clogs up the animals' stomachs and intestines.

Every year more than 1,000 people volunteer to pick up over 150 tons of trash on Padre Island. Volunteers also clean up other beaches. When you go to the beach, remember to pick up your own trash. You may also want to join others at your favorite beach on volunteer cleanup days or just do it yourself.

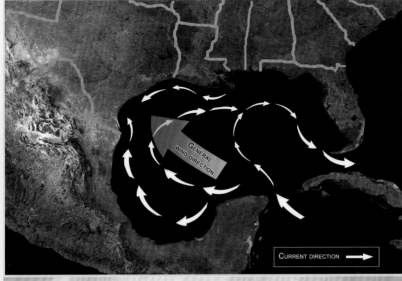

Map courtesy of Harte Research Institute for Gulf of Mexico Studies and modified by Rudolph Rosen (*top*); photograph by Rudolph Rosen (*middle*); photograph courtesy of *Corpus Christi Caller-Times* (*bottom*).

AQUATIC SCIENCE CAREER

Marine Biologist

Marine biologists are scientists who study saltwater organisms and their relationship to the marine environment. They work in the ocean, coastal wetlands, seashores, estuaries, or anywhere else there is saltwater. They may work on a boat, scuba dive, use a submersible vessel, or study marine life from shore. Some marine biologists work in a laboratory, where they may examine tiny creatures under a microscope; on a large research vessel; or even at a marine life educational park and aquarium, where they may work with dolphins and whales, fish, seabirds, invertebrates, and many other animals in a giant marine tank at an aquarium. Marine biologists have at least a bachelor's degree, and many also have a master's or doctorate degree that gives them added options for exciting employment opportunities.

Photograph courtesy of Texas Parks and Wildlife Department.

chapter 13

Fishing for Conservation

Questions to Consider

- What does it mean to think like a fish?
- How can knowing about aquatic communities and food webs be used to improve fishing success?
- How can knowledge of fish adaptations be used to improve fishing success?
- What is "cover" for fish, and why is it important in fishing? What weather factors improve fishing success?
- What are at least five observations that would help an angler be successful?
- Why are rules about how many fish you are allowed to catch important?
- What is an ethical angler? What are some important things an ethical angler does?
- How do anglers contribute to fish conservation?

CHALLENGE QUESTION

What are some things you can do as an angler to improve aquatic ecosystems for the future?

Now that you have learned about the aquatic ecosystems in Texas, one of the most fun ways to use your knowledge is to go fishing, which is a great way to spend time with friends and family outdoors. With over 191,000 miles of rivers, 212 major lakes, 367 miles of coastline, and 3,300 miles of bays and estuaries, Texas has many places to go fishing. Hike or take a canoe or boat to a fishing spot. Try camping near a lake. Cook what you catch and have a picnic. Fishing is a great way to learn about nature. When you go fishing for fun using a rod and reel, you are called an **angler** or a **sport fisherman**. To be a good angler, you need patience, fishing skills, and knowledge about aquatic ecosystems.

Think like a Fish to Catch a Fish

You can use what you have learned about fish, habitats, food webs, niches, trophic levels, and aquatic ecosystems to help improve your fishing success. Fish may scour the bottom, hunt near the surface, or swim anywhere in between. Spawning brings fish together in one place. Their need for cover, where fish hide, attracts them to structures in the water, such as rocks, logs, and aquatic plant beds (fig. 13.1). They seek comfortable temperatures because they do not use energy to maintain a constant body temperature as do warm-blooded animals. They also may move about to avoid low oxygen levels. Individual and species needs and preferences present a constantly changing yet somewhat predictable pattern of fish activity.

To catch a fish, it helps to think like a fish. What habitat does the fish like within the aquatic ecosystem? A catfish stays near the bottom and may hide inside or under objects. A largemouth bass prefers to swim in midwater and to be near the surface but may hide in cover so it can ambush its prey. What kind of food does the fish you are trying to catch like to eat? Use the fish's

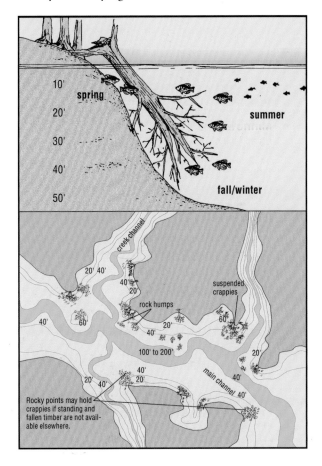

Figure 13.1. Many fish species that are fun to catch, such as crappies, use cover habitat and structure. Which cover and structure are used most may vary based on season, the type and amount of cover, and depth or location of structure, such as a large tree in the water. Illustration courtesy of In-Fisherman.

natural food as bait, or use an artificial bait that looks or smells like something the fish would normally eat (fig. 13.2).

Fish tend to gather where there is plenty of the kind of food they like. Look for this food. Schools of minnows or other prey fish in a lake will attract larger open-water fish to feed on them (fig. 13.3). Sunfish watch for hatching insects. Largemouth bass search for frogs jumping in a pond. You should watch for this, too. Watch for small fish dashing about the surface of the water in bays and estuaries. Even the sound of a frog jumping or a small fish splashing on the surface can attract predator fish. Birds eating small fish on

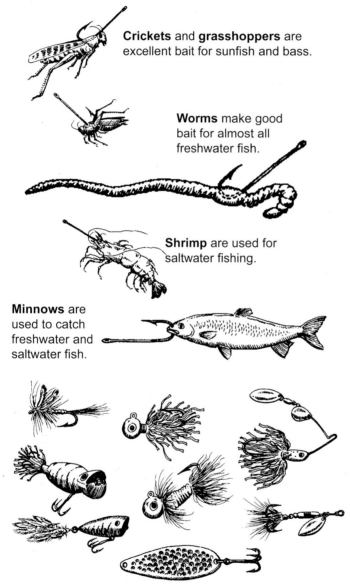

Figure 13.2. Popular fishing baits. Illustrations courtesy of Texas Parks and Wildlife Department; modified by Rudolph Rosen.

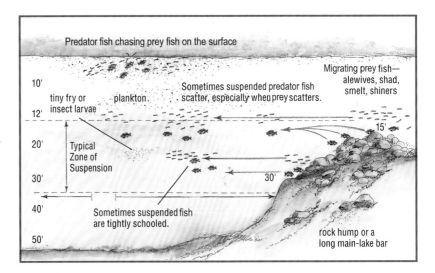

Figure 13.3. Some species of fish may move off a structure and into open water to feed, where they suspend themselves in the water. This is often in response to following or searching for prey. You are most likely to find fish where their food is available. Illustration courtesy of In-Fisherman.

the surface are another clue. All these sights and sounds can lead you to fishing hot spots. Small fish darting above the surface are often trying to escape from larger fish, such as red drum. Look for the signs of feeding fish, or even fish eating birds. Rings spreading across the surface of a pond could mean that bluegill are feeding on insects. A splash in the shallows could be a largemouth bass chasing a frog. Cast your bait where you think fish are feeding.

Finding Fish

Fish use cover to escape predators and to help them ambush prey. Ask yourself, "If I were a fish, where could I hide from enemies and find food?" Cover can be anything that will hide or protect the fish. Some fish spend most of their lives near cover, moving out only to feed or spawn. Aquatic plants, docks, rocks, undercut banks, coral reefs, and logs all provide cover. Something like the shade, or shadow, from a tree can be cover, too. It makes the fish less visible to other fish, predatory birds, wildlife, and even people. Look under overhanging trees, docks, or fishing piers.

Once you start looking, you will see all kinds of cover and maybe see some fish. Weeds grow near the bank, fallen trees lean over the water, fishing piers reach out into the bays and Gulf, boat docks rim lakes, and big rocks often rest in the middle of streams. In the Gulf, oil rigs and sunken objects such as boats provide fish habitat deep underwater and are great spots for fishing (fig. 13.4). As in the Gulf, a lot of cover for fish cannot be seen from the surface. Underwater rocks and sunken logs rest on many lake and river bottoms. Cast close to cover because that is where the fish hide out.

Drop-offs, points, ridges, islands, and sandbars shape the beds of lakes, rivers, bays, and the Gulf. These features often attract more fish than do flat or gently sloping bottoms. You can find good places to fish from clues on land or in the water. Land points often extend into a lake (fig. 13.5). The open area between flooded trees might be an old river channel. Texas bays and estuaries

Figure 13.4. You need a big boat and knowledge of the location of oil rigs in the Gulf to find some of the best saltwater fishing in Texas. If you do not have a boat, you can go on a charter fishing boat for a day of fishing in the Gulf. Photographs courtesy of Texas Parks and Wildlife Department.

have oyster reefs that provide fish cover. A change in the wave pattern in the bay may signal the edge of a seagrass bed.

Hungry fish seem to like places where one kind of habitat changes to another—in other words, at the border or "edge" between habitats. The edge of a lake's shoreline area, or littoral zone, for example, usually attracts many fish (fig. 13.6). In rivers, fish often feed where the flow changes direction or slows down. You can often see signs of these changes from shore or from a boat. Look for the break between muddy and clear water in rivers and estuaries.

In flowing water, there is less current near the bottom or immediately downstream from an object like a rock. Because of the slower current, most stream fish rest with their bellies almost touching the bottom or stay in the slower water downstream of objects. They like to take advantage of spots that

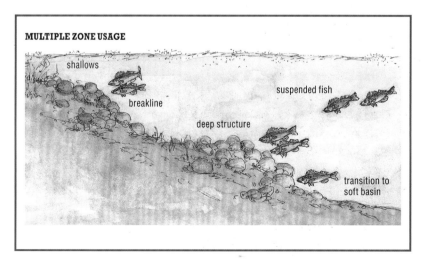

Figure 13.5. Many large bodies of water offer multiple opportunities for different groups of fish of the same species to inhabit different lake zones simultaneously. Individual groups of fish may or may not interact and mix. Fish inhabiting different zones tend to become active at different times. They may be active in shallow zones in the morning, in the evening, and at night. In deep zones they may be most active during periods of brightest sunlight. Each individual group may tend to stay more or less within a defined depth range rather than make daily movements up and down a structure. The more you learn about the structure available to the fish and zero in on it, the more fish you will catch. Illustration courtesy of The In-Fisherman.

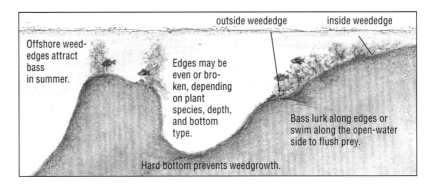

Figure 13.6. Fish are usually caught near cover along the edges of habitats in lakes, rivers, and bays. You can often see these places from shore or a boat. The edge of underwater plant or seagrass beds, for example, usually attracts many fish. Illustration courtesy of In-Fisherman.

have less current than the nearby water. They do this to save their energy and to avoid being pushed downstream. The faster water brings food past waiting fish looking for a quick bite to eat. Most fish in a river face upstream into the flow of water and patiently wait for food to come to them. Cast your bait in front of fish in a stream and let the current take it to the hungry fish.

Go Fishing

Fishing can be good at any time. However, fish seem to prefer eating during the low-light conditions of morning and evening rather than in the bright sun of midday. Fish stay in shallower water in low light and then move to deeper water when the sun is bright. Cloud cover or shadows on the water mimic low-light periods and may help get fish to bite. Catfish, bass, crappie, and many other species of fish will bite day or night. In some clear lakes, fishing is sometimes better at night than during the day. Big fish seem to be less picky about food and are easier to catch when it is dark. Fishing at night is hard, even for experienced anglers. If you are just starting out, try fishing during the evening and see if you can continue after dark.

Weather affects fish but not always in predictable ways. Hotter weather sends fish deeper to find cooler temperatures, such as in the open-water zone of a lake. Most fish stay out of the deep-water zone of the lake because there is little oxygen there. Warm fronts in winter may improve fishing, and the longer the front stays, the better. Cold fronts may reduce fish movements, so the fishing is not as good. Many people agree fishing is affected by weather changes, for better or worse. A light to moderate wind is better than no wind. A little wind gives fish in shallow water cover from fish-eating birds, because the fish are harder to see from above when the water is rippled by wind. Fish will move into shallower water to feed in slightly windy conditions. Fishing can be good before and during a gentle rain but may be poor during and after a big thunderstorm. If this all seems like a lot to learn, just remember that anytime you can safely fish is a good time to go fishing, and you will have fun even if you do not know everything right away.

Fishing Rules Make Fishing Better

The Texas Parks and Wildlife Department (TPWD) makes and enforces rules to conserve fish and wildlife. These rules help Texans share and keep our aquatic resources healthy. Fishing rules protect species by limiting the number or size of fish that may be taken. Length limits give fish a chance to grow and spawn before people are allowed to catch and keep them (fig. 13.7). "Slot" limits are restrictions on keeping fish of varying lengths. These kinds of rules allow biologists to achieve very specific purposes. A slot limit may be used to allow larger, very productive fish to stay in the habitat and produce more offspring for anglers to catch. For example, red drum have a 20- to 28-inch slot limit. You are allowed to keep fish within this size range, but you must return all fish that are smaller or larger. These regulations make fishing better, but they also require anglers to know the rules and carefully measure the fish they want to keep to make sure they are the legal size.

Texas' fishing rules are based on scientific data and research provided by fisheries biologists. Taxes on fishing equipment and the sale of fishing licenses pay for this work. When you turn 17 years old, you are able to buy a fishing license and help support fish and fish habitats in Texas. If you are too young to buy a license or if you go fishing where a fishing license is not required, you must still obey fishing regulations. This way everyone plays a role in fish conservation. When you purchase a fishing license or fishing equipment, you are supporting fishery management, hatcheries, conservation, and education (fig. 13.8). By learning to identify fish and respecting fishing regulations, anglers help keep fish populations in balance. Catch and release is a good conservation practice, but harvesting some fish can give others room to grow.

So before you go fishing, you need to learn the rules and follow them. The rules can be found in the *Outdoor Annual* published by TPWD every year and free anywhere licenses are sold. The rules can also be found on the TPWD website. Following fishing rules and being an **ethical** angler are ways you can help conserve the aquatic ecosystems in Texas. Conservation helps make sure our aquatic ecosystems stay healthy far into the future.

Figure 13.7 (*left*). Anglers measure fish to make sure they keep only the ones that are of legal size. Photograph courtesy of *Corpus Christi Caller-Times*.

Figure 13.8 (*right*). Anglers make it possible for Texas Parks and Wildlife Department biologists to better manage all the state's fish and fishing opportunities. Money from the sale of fishing licenses and a special tax on fishing equipment and gas for boats help pay for kids' fishing programs, fish hatcheries, surveys of fish in lakes and the ocean, and many more things that make fishing in Texas better. Photograph by Rudolph Rosen.

BE AN ETHICAL ANGLER

Do the right thing. When making a choice, ask yourself the following questions:

- Is it legal?
- Would it be good if everyone did it?
- Would it make you proud?

Ethical behavior is more than just following fishing regulations. Ethical anglers go beyond what laws require and help others understand and obey the rules. Ethical people use good judgment even when nobody else is watching. They respect property rights and never go on private lands or waters without permission.

Ethical anglers respect the rights of others who are fishing or using the water. Ethical behavior includes picking up your trash, collecting and properly disposing of any fishing line you find, and never dumping any pollutants, such as gasoline or oil, into the water. The ethical angler values and respects the aquatic environment and all living things in it. Never release live bait fish into the water because this can introduce unwanted, potentially harmful, species. Carefully handle and release alive all fish you catch but do not intend to keep.

Photograph courtesy of Texas Parks and Wildlife Department.

Fishing Is Good in Texas

Texas Parks and Wildlife Department recognizes fishing excellence through its Angler Recognition Program. The program maintains records of biggest fish caught in public and private waters and records of biggest fish from all public lakes, rivers, and bays. Junior Anglers (under 17 years of age) com-

Figure 13.9. If you quickly remove the hook, it is safe to have a friend take a quick photo of you and your fish before releasing the fish back into the water. Photograph courtesy of *Corpus Christi Caller-Times*.

pete in a separate division. For more information, go to the TPWD website (http://www.tpwd.texas.gov/).

Texas has many **community fishing lakes** where anyone can fish. These are public lakes 75 acres or smaller located in city, community, and state parks. These lakes are stocked annually with fish such as channel catfish and rainbow trout for fishing. Information about fishing and stocking in community fishing lakes and over 100 major lakes in Texas can be found on the TPWD website.

You can also go to the website to learn more about conserving Texas' aquatic resources. Better yet, go outside and visit your favorite local aquatic resource. Begin thinking of it as *your* lake, pond, river, stream, or bay. To help keep our lakes beautiful, consider bringing a trash bag when you visit, and take a moment to leave the spot in better shape than you found it.

There are many ways you can help keep fishing great in Texas. The simplest way is to just start fishing! Learn to recognize and report invasive aquatic species that can take over and ruin native habitats and area fisheries. Be safe and responsible on the water. Be a mentor and take someone fishing! Visit a fish hatchery. Join Texas Stream Team to learn more about water and become active in conserving water quality. Join a local fishing club. There are even fish conservation organizations that hold fishing contests where winners get college scholarships (fig. 13.9). And if you are up to the challenge, choose a career in conservation and make aquatic resources your life's work. Above all, enjoy your aquatic resources and use them wisely!

TIPS FOR RELEASING FISH SAFELY

1. As a safety measure for you and the fish, mash down the barb of your hooks with pliers before you go fishing.
2. Quickly catch and then release fish as soon as possible.
3. Before handling a fish, wet hands to help avoid removing the fish's protective slime.
4. Take photos quickly; remember that the longer the fish is out of water, the more stressed it gets.
5. Remove the hook from the fish with pliers, or cut the line if the hook has been swallowed.
6. Gently place the fish back into water, taking care not to bump it against the ground or your boat.
7. If the fish appears sluggish or dazed, gently try to revive the fish by holding it upright in the water and moving it back and forth, slowly forcing water through its gills.
8. If you do not intend to eat a fish, do not put it on a stringer.

Photograph courtesy of *Corpus Christi Caller-Times*.

ENVIRONMENTAL CONTAMINANTS IN FISH

The Texas Department of State Health Services (TDSHS) monitors fish for the presence of environmental contaminants and alerts the public through bans (closures) and advisories when a threat to human health may occur from eating contaminated fish. In waters with consumption bans, possession and consumption of fish and/or shellfish are prohibited. A consumption advisory is a recommendation to limit consumption to specified quantities, species, and sizes of fish. For more information, visit the TDSHS website (http://www.dshs.state.tx.us/).

Photograph courtesy of Texas Parks and Wildlife Department.

AQUATIC SCIENCE CAREER

Fish and Wildlife Conservation Officer (Game Warden)

Conservation officers enforce laws that control the use of our fish and wildlife. They make sure that people who hunt or fish for recreation or who take fish or wildlife to sell do so according to all the laws. This includes making sure people have proper licenses. These officers also work to protect endangered species and make sure that people do not pollute our waters. Conservation officers work for local, state, or federal government agencies. Most conservation officers in Texas have a bachelor's degree and special law enforcement training.

Photograph courtesy of Texas Parks and Wildlife Department.

chapter 14

Online Video under ch. 13

Water for People and the Environment

Questions to Consider

- What is the purpose of the Clean Water Act? What kind of water pollution is it intended to prevent? Why does the act not stop all water pollution?
- What is watershed action planning? What should people where you live consider when conducting watershed action planning?
- In Texas, who owns surface water? Who owns groundwater? Since all water is connected, why do we have different systems of regulation?
- What is a water right? What are the impacts of prior allocation?
- What is rule of capture? What rights and responsibilities are associated with it?
- What are environmental flows? How are they obtained?
- What will influence Texas' water supply for the future?

CHALLENGE QUESTION

In your opinion, should the future of Texas' water quantity and quality be driven by laws or by voluntary action? Defend your answer.

As you have seen throughout this book, we are all connected by water. All living things need water. The Earth has a remarkable system of transporting and recycling water for our use. Because we live on a planet with a finite supply of water, we must take care of our natural resources, especially the water we use where we live. Whether we are a community of humans or a community of aquatic plants and animals, we all live downstream from someone or some organism that needs and uses the water. We should treat the water that we use the same way we hope the water we use was treated by others.

To a large extent in Texas, water quality and quantity within a watershed are linked to the actions of the people who live, work, and play within its boundaries. Some laws exist to make sure we, and aquatic life, have adequate clean water. Water gets polluted or used up in some ways that are not the subject of strong laws. In these cases, education, use of science, best practices, a conservation ethic, and personal voluntary choices and actions are needed to ensure water sustainability for Texas.

Clean Water Is the Law!

In 1972 the US Congress passed the Federal Water Pollution Control Act, better known as the **Clean Water Act**. The goal of the act is to make surface waters fishable and swimmable. The act made it unlawful to release any pollutant from a point source into most major US waters without obtaining a permit to do so. It also made filling of wetlands unlawful without a permit. The act does not include waters that do not connect to navigable waters, although not everyone agrees which waters are exempt. The Texas Commission on Environmental Quality monitors **pollution** and issues permits for discharge of **pollutants** in Texas.

Because nonpoint source pollution comes from many places, it is difficult to trace to its source and therefore is not controlled by the Clean Water Act. A common cause of nonpoint source pollution is runoff, such as when rain or floodwaters carry pollution as it runs off the land. This kind of pollution is blamed for high amounts of nitrates in groundwater and sediment in surface water, which are the greatest contaminants in many Texas waters today. Agricultural activity accounts for most of this nonpoint source pollution. The irrigation and plowing of crops, and application of fertilizers and chemicals, generate pollutants such as nutrients, pesticides, insecticides, herbicides, and sediment.

Nonpoint source pollution comes from urban areas as well. This includes fertilizers, herbicides, and insecticides from lawns; sediment from construction sites; and oil, grease, and toxic chemicals from roads and parking lots. We all contribute to this kind of pollution when we drive our car or fertilize our lawn. Because of its diffuse nature, nonpoint source pollution can be more difficult and costly to describe and control than point source pollution. This is one reason there is no single law and permitting system for nonpoint source pollution as there is for point sources.

Responsibility for nonpoint source pollution control in Texas for agriculture and forestry falls on the Texas State Soil and Water Conservation Board. The Texas Commission on Environmental Quality is responsible for control of all other nonpoint sources, such as runoff from urban areas. Texas also relies on education, water quality management programs, and volun-

Figure 14.1. Texas Stream Team is a statewide network of citizen monitors who volunteer to test water quality. Volunteers include teachers and their students. The volunteers monitor water quality and report results for addition to a statewide base of information about the condition of Texas' streams, rivers, and other water bodies. Students monitor a body of water under a teacher's supervision, with activities based on the educational objectives of the class. Certified teachers can also train students to become volunteer monitors who can then go on to form groups and monitor their own aquatic sites. Students who measure what is happening in the aquatic environment develop an understanding of water quality issues. They are better prepared to participate in local water resource issues as well as make environmental work a career choice. Photograph courtesy of The Meadows Center for Water and the Environment.

tary, financial, and technical assistance to reduce nonpoint source pollution. Another way Texans take care of water is through watershed action (protection) planning. Landowners, scientists, local government officials, and staff from responsible agencies consider sound science to design, implement, and monitor water quality management strategies to protect and restore water quality (fig. 14.1).

Texas Water Law

Texas water law regulates surface water and groundwater differently. Surface water belongs to the state, but groundwater belongs to the landowner. History plays a big role in the development of Texas water law. The legal rights to own and use water in Texas came from the cultures and legal systems of Mexico and England. Early Spanish settlers in Texas used extensive systems of ditches to move water from place to place and to irrigate their crops. These early water systems were managed by the communities for the people served. This centralized control of water was generally based on Spanish law and is the origin of Texas' system of prior appropriation.

WHO DOES WHAT REGARDING TEXAS WATER

Many government agencies, local authorities, and private companies an organizations work on water and aquatic environments in Texas.

- The Texas Commission on Environmental Quality and US Environmental Protection Agency are responsible for monitoring water quality and for setting and enforcing water quality standards.
- The US Army Corps of Engineers and various river authorities, such as the Trinity River Authority and Lower Colorado River Authority, control river flows and manage large reservoirs.
- The Texas Water Development Board and groundwater conservation districts, such as the Red River Groundwater Conservation District and the Uvalde County Underground Water Conservation District, develop, approve, and implement groundwater management plans.
- Many other agencies have a say in matters affecting soil, water, and wildlife. These include the US Fish and Wildlife Service, the Texas State Soil and Water Conservation Board, the National Resource Conservation Service, the Texas General Land Office, the Texas Forest Service, and the US Forest Service.
- Texas Parks and Wildlife Department has no direct regulatory authority over water but has an indirect role and is responsible for protecting aquatic species. The agency's biologists and other conservation professionals are in the forefront of developing environmental water flows and water quality strategies for streams, rivers, bays and estuaries, and the Gulf. They help recover endangered and threatened species, control the spread of invasive species, regulate fishing and wildlife use, investigate fish kills and pollution spills, and ensure that Texas has adequate supplies of clean water for people, fish, and wildlife.

Photograph courtesy of Texas Parks and Wildlife Department.

Other people help address water right matters, as do regional water and watershed planning groups, water authorities, local water supply districts, drainage districts, and navigation districts. People may join local, regional, or national conservation organizations that work on water matters, including aquatic habitat and species protection. The members of these organizations often raise money for protection projects and advocate for sound conservation policies and laws before state and local elected officials and the US Congress.

Prior appropriation gives the right to a certain amount of water to certain users. It divides available water among people who want it based on a government-issued permit that includes the principle of first come, first served. Senior water rights holders have first rights to a prescribed amount of whatever amount of water is available. This is also often referred to as the "first-in-time, first-in-right" rule.

Anglo-American settlers who moved into Texas in the early 1800s brought with them a different system granting rights to water called riparian law. This gives people who own land bordering streams the right to use water from the stream. It is based on English law and reflects conditions in England where, unlike in Texas, water is plentiful and there are many streams everywhere.

Over time, these systems were merged and Texas became a **dual doctrine** state, recognizing both riparian and prior-appropriation rights. Neither of the two systems of allocating water considers ecological needs of aquatic ecosystems or the direct connections between groundwater and surface water. Today, Texas courts divide water into unrelated legal classes with different rules governing who can own and use water in the different classes.

Classes of Water Allocation

Texas water law can be thought of as being contained in three buckets, depending upon the geologic location of the water. There is a bucket full of laws for natural surface water, one for groundwater, and one for diffused surface water.

Natural Surface Water

Natural surface water found in waterways, such as rivers and streams, is owned by the State of Texas and managed for the citizens of the state. This includes the ordinary flow in streams and tidal waters on the coast. Water from rains and floodwater found within natural rivers, streams, and lakes and in human-made reservoirs on waterways is also state water. Water in springs that form headwaters is also considered surface water.

Surface water is allocated to people through a water right permit. These permits give holders a set amount of water on an annual basis from a specific location on the water body. Once people have a water right permit, they can sell all or part of the water right to other people. Because it is not possible to say exactly how much water will be in a stream, lake, or spring at any point in time, water is allocated among the permit holders according to who received rights to the water first. This is the system of prior appropriation.

A water right is like a ticket for a place in line for available water. In times of drought when water in streams and lakes may be very low, the person who holds the first water right permit may take the full amount allowed first from the water available. The next most senior permit holder may then take the allowed full amount, and so on. Some people who have permits for water may not receive any water in drought years. In some Texas streams, more water has been allocated than flows in them in dry years.

Groundwater

Texas treats groundwater, the water in our aquifers, and surface water differently. Whereas surface water is considered state property, the water beneath a landowner's property is private property. Landowners have a legal right to pump and capture whatever groundwater is available, regardless of the effect of that pumping on neighbors' wells or springs that may be fed by the groundwater. This is called **rule of capture**.

Since groundwater exists underground, the surface landowners do not actually own the water until they take it from the ground, or capture it. Once the water is captured, landowners have the right to use the water or sell it to others. The right to capture the water can even be sold to others. Once sold or leased, any water captured may be transported by the new owners wherever they want. There are some limits, however. Landowners are not to pump an unlimited amount of water when it is done maliciously to harm a neighbor, in a wasteful manner, or through negligence that causes **subsidence** in a neighbor's land.

The amount of groundwater a landowner can take may also be restricted when the water is from an underground river or is regulated by a groundwater conservation district. There are about 100 groundwater conservation districts now in Texas, formed to create rules for conserving, protecting, recharging, and preventing waste of underground water. Some exert more control over use of groundwater than others. They can register and permit wells, keep drilling and well records, regulate how far apart wells are drilled, require a permit for water transfers, buy and sell water, and generally conserve and protect the aquifer. Despite these ways to conserve groundwater, many of Texas' aquifers are being depleted, which affects future water supplies (fig. 14.2).

Figure 14.2. Overuse of groundwater impacts Texas' rivers and streams. A spring known as Jacob's Well in the Hill Country is the headwaters of Cypress Creek, which merges with the Blanco River and eventually flows into San Antonio Bay. Jacob's Well had never stopped flowing until the last 10 years. But it has now gone dry twice due to increased groundwater use from population growth in the area. Without conservation, the well could go completely dry. Comanche Springs in Fort Stockton once fed water to a 30-mile-long stream. As a result of groundwater pumping, the spring has stopped flowing, except for brief periods after heavy rains. Photograph courtesy of Texas Parks and Wildlife Department.

Diffused Surface Water

Water such as runoff from rain that does not flow in a defined watercourse, but flows generally across the surface of the land, is diffused surface water. Generally, this is rain runoff or floodwater left in upland areas after the flood recedes. This surface water belongs to the landowner until it enters a natural watercourse. Landowners can use the water however they wish. When it enters a natural watercourse, it becomes state water.

Riparian Rights

Texas also has remnants of riparian law. For example, landowners whose property borders a stream can take up to 200 acre-feet of water in a year for domestic and livestock use anytime they want. This is just for a home, farm animals, and gardens, not for big commercial farms or ranch operations. Up to 200 acre-feet per year can also be impounded behind a dam on your property. One **acre-foot** is enough water to flood an entire football field to a depth of nine inches.

Water for the Environment

While in the past many people did not think about leaving enough water in a stream or river to support its aquatic life, there are new laws in Texas that require **environmental flow**. Also called instream flow, this is an amount of freshwater, or flow, left in a river or other water body that is adequate to support an ecologically sound aquatic environment.

Allocating this amount of water to many streams in Texas can be complicated. In a number of river basins, permits have been issued for all, or even more than all, of the water that is present during dry periods. If all water rights were fully used, those rivers and streams would go dry during a drought, leaving nothing for fish and wildlife. This would cause problems for freshwater species all the way downstream and into the estuary and bay, where freshwater is important to saltwater ecosystems.

New laws require that some water be set aside for environmental flows in areas where water rights permits are still available. Where all the water has already been allocated to someone, environmental water will need to be acquired over time. For example, existing water rights owned by others could be donated or purchased and allocated to environmental flows. We could also improve ways to use water efficiently, leaving more water to return to our water bodies.

Figuring out how much water to leave in a stream to maintain aquatic life from the headwaters all the way to the bay is very complicated. The amount of water in a stream naturally varies by season, by location along the river, and by year. Different species and habitats need varying amounts of flow at different times. But the most critical time to ensure adequate environmental flows for aquatic species is during periods of drought.

To help find out how much water is needed and how to get it, scientists and **stakeholders** come together to find answers. The streams, rivers, reservoirs, estuaries, and bays in each Texas basin are unique and have their own requirements for environmental flows. For study and planning, separate groups of scientists and stakeholders formed for 11 river basins. Scientists conduct studies to determine how much freshwater should be left in streams at any particular location, as well as how much should flow out into the estuaries and bays to meet the needs of fish, wildlife, aquatic ecosystems, and people.

Stakeholders are people who have an interest in water allocation and who live in the river basins. They are informed of scientific studies and come up with ways to protect and acquire more water for the environment. They make suggestions that include water conservation and efficient use incentives, use of treated wastewater, and the purchase or donation of existing water rights.

The Water Future of Texas

How much water Texas will have for the future will depend on what we do to conserve water today. We cannot make it rain more, but we can more efficiently manage the water we get from the hydrologic cycle and water in our aquifers. We may have enough water for people to drink. But will we have enough to sprinkle on our lawns, wash our cars, irrigate our crops, or remain in our streams to support aquatic habitats and flow into our bays and estuaries?

Our primary users of water are farmers and ranchers who grow our food; workers in businesses and industry who make and sell our products; and all of us who live in cities and towns and use water for drinking, washing, and watering our lawns, gardens, and parks. Improved irrigation methods and agricultural practices have already reduced the water needed by many farmers and ranchers. Conservation measures and repair of leaking municipal water lines have helped some cities such as San Antonio reduce their water demand considerably. But with an increasing population needing jobs and places to live, the demand for more water for food, industry, and people will continue to grow.

The human population in Texas is expected to almost double from 25 million in the year 2010 to more than 46 million by the year 2060. All these new people will need places to live, more roadways and industry, and more parking lots. There will be more people drinking water, bathing, and using water in their businesses. Demand for water in cities will grow. Without wise use of water and responsible conservation practices, Texas will have increased nonpoint source pollution, less groundwater recharge, more erosion, reduced environmental flows for aquatic ecosystems, and less freshwater inflow to estuaries, which will reduce production of seafood. More growth could also mean more loss of wetlands, riparian areas, and stream buffers, which would reduce the environment's natural ability to absorb floodwaters and to filter contaminants before they flow into our bodies of water (fig. 14.3).

Figure 14.3. Without conservation of Texas' waters, including providing adequate freshwater inflow to our bays and estuaries, catching redfish like these will become a thing of the past. Photograph courtesy of Texas Parks and Wildlife Department.

Texans have shown an ability to wisely manage water use, and there are many examples in our cities and on private lands of how to do so, but have we waited too long? Of 281 springs that were identified as historically significant, more than 63 have stopped flowing since the 1800s. We have lost over half of our wetlands. More than 25% of our native freshwater fish species are considered at risk, and we are in the top five states for number of endangered aquatic species.

Texans will need to commit to more intensive water conservation and pollution prevention practices to reach a sustainable water goal that also protects aquatic ecosystems. It will take an understanding of water, watersheds, and life in Texas' many aquatic environments. It will obligate us to reserve water for species and habitats from our headwaters to ocean. It will require the combined efforts of our leaders in business and industry, our elected officials, our teachers, and you.

PEOPLE & THE ENVIRONMENT

YOU CAN MAKE A DIFFERENCE

Do you believe that everyone deserves a sustainable and adequate supply of clean, safe water for our homes, farms, and industries? Do you believe fish, wildlife, and all other aquatic life need an adequate supply of clean water, too? If so, you can help ensure this happens in Texas. Here are ways you can help make a difference, as a student and as an adult. You may be able to think of other ways to help where you live.

- Learn where your drinking water comes from and tell others.
- Become a volunteer water quality monitor through the Texas Stream Team, or have your entire class monitor water quality.
- Learn about water conservation measures you can take and ways you can reduce pollution where you live.
- Help rescue stranded marine mammals, for example, volunteer through the Texas Marine Mammal Stranding Network.
- Join a conservation organization that has a mission of conserving water resources or aquatic species. There are many to choose from, some of which are formed to protect local water resources, such as a particular spring or river.
- Help restore an oyster reef by recycling shucked oyster shells.
- Buy a federal Duck Stamp because money from the sale of Duck Stamps is used to protect and restore wetlands important to waterbirds.
- Volunteer to pick up trash from the beach or from around your local lake or stream. For example, volunteer through Texas Adopt-A-Beach.
- Become active in the public debate on water and aquatic ecosystem conservation, and attend and comment at public hearings.
- Buy a fishing license and go fishing, and even if you do not go fishing, funds from licenses are used to manage the state's aquatic resources.

Student volunteers help with a Kemp's ridley sea turtle release by the National Park Service on Padre Island. Photographs by Jennifer Idol, The Underwater Designer.

AQUATIC SCIENCE CAREER

Park Ranger

Park rangers work to educate people about the natural environment, history, and cultural resources of a park, natural area, forest, or other area of land or property. They also work to protect and conserve the natural environment, historic buildings, geologic features, cultural resources, and anything else of importance on the site. They help visitors, create recreational opportunities, and ensure visitors do not harm any resources. Duties are varied and include talks and tours for visitors about history, geology, local wildlife, or ecosystem science; arts and crafts demonstrations; control of traffic and visitor use of facilities; enforcement of laws and regulations; fire prevention and control; investigation of violations and complaints; search and rescue missions; and management and protection activities related to resources such as fish, wildlife, lakeshores, seashores, forests, historic buildings, archeological sites, and recreation areas.

Park rangers work for local, regional, state, and national governments and can also work for private organizations and corporations that conserve land and properties. Education level ranges from high school diploma to a master's degree and above. Degrees can be in many areas, as park rangers' duties cover so many areas from geology to biology, history to law enforcement, paleontology to blacksmithing, and much more.

Photograph courtesy Rudolph Rosen.

PEOPLE & THE ENVIRONMENT

Appendix A

Guidelines for Teachers

Texas Aquatic Science is a comprehensive aquatic science curriculum consisting of this book, a teacher guide, specially produced videos, and supplementary materials. The following describes the various components of the curriculum and provides suggestions on use for teachers. The teacher guide, videos, and other resources can be found at http://texasaquaticscience.org/ and at http://www.tpwd.texas.gov/publications/learning/aquaticscience/.

The Book

Texas Aquatic Science provides clear, concise scientific information about water and aquatic life from the molecular level to the level of aquatic ecosystems, spanning the hydrologic cycle from watersheds to aquifers to headwaters to ocean. Illustrations help readers understand important concepts and clarify major ideas. Aquatic science is covered comprehensively, with relevant principles of chemistry, physics, geology, geography, and biology included throughout the text. Readers are introduced to concepts of water sustainability and conservation of ecosystems and species, including what they can do personally to conserve water and aquatic resources for the future. Each chapter begins with a series of questions, which help readers know what to focus on in their reading and link to lessons in the teacher guide.

Readers are also introduced to careers in water sciences and conservation through a series of job descriptions and photographs of young professionals working in these careers. There are also stories about interesting aquatic species, amazing concepts in biology, wonders of ocean dynamics, and protection of natural resources. Readers learn how they can volunteer to help restore species to aquatic environments. Job and volunteer opportunities are illustrated in the book in the hope that some readers will become excited by aquatic science and pursue a career or become a volunteer that involves them in helping build a sustainable future for our aquatic resources.

Teacher Guide

Accompanying the *Texas Aquatic Science* textbook is a comprehensive teacher guide, available to educators for free download at http://www.tpwd.state.tx.us/publications/learning/aquaticscience/. With this guide, teachers can introduce students to the wide variety of aquatic ecosystems through science investigations, games, models, cooperative learning activities, Internet projects, readings from the website and/or book, science journals, and field-based assessments of water quality and environmental conditions in a variety of field trips.

The guide is divided into chapters and provides activities that correspond to each chapter in the textbook. The guide is also linked to short videos that provide an overview of the main ideas in each chapter. Together the guide, book, and videos lead students to an understanding of the characteristics of water that make it unique and essential to all life.

Lessons

Lessons in each chapter of the teacher guide begin with an activity to allow the teacher to assess what students know about the concepts to be studied. Lessons embed higher-order thinking skills, provide depth and complexity of learning, and offer hands-on activities that provide various contexts and methods for the student. Lessons include a variety of components. Students use science journals, participate in cooperative learning activities, take part in different learning evaluations, and collect data through field investigations.

Cooperative learning activities are included throughout the lessons. Some activities are as simple as designing and conducting investigations in small groups, where students each have a part in making the work go smoothly. Other activities include a variety of ways for students to help each other by breaking down concepts into parts and teaching what they learn to the rest of the group. Many activities involve students brainstorming concepts or questions.

Each lesson includes an opportunity for students to apply and demonstrate what they have learned by developing creative products or performances. Activities are designed to be inexpensive and to use materials that are already in most classrooms. Teachers will find the activities easy to implement and fun for students. Teaching materials for some lessons, such as aquatic organism game cards, posters, and videos, are contained in the teacher guide or are available for teachers to download or view.

Science Journals

Keeping a science journal provides opportunities for students to record their discoveries, questions, experiments, observations, reflections, labeled drawings or diagrams, data tables, and graphs. Systematic records of their work help students develop awareness and understanding of their experiences. Writing down what they see and do helps them put learning into words, and having the written record helps them review and think about their learning. Science journals also increase the teacher's awareness of student progress in science process skills and concepts. The journal then becomes a tool for helping parents understand the student's development over time.

Field Investigations

The teacher guide includes eight field investigations, which are trips to the outdoors that vary in purpose and destination. Information on group management and safety are included. Teachers may choose to do one or all of the field investigations depending on time, transportation, and financial constraints. Field investigations introduce students to a local creek, pond, or estuary; to water chemistry testing; to a scavenger hunt looking for places where water flows, infiltrates, or accumulates; and to techniques for estimating the population numbers and composition of aquatic species.

Assessment

Each chapter of the guide provides multiple opportunities for assessment. The first lesson in each chapter provides a formative assessment to help teachers plan for appropriate student learning and to help students focus on what is to come. In addition, every lesson has a component that allows students the opportunity to incorporate what they have learned in creative exercises. The student science journals are also useful for formative and summative assessments.

At the end of each chapter of the teacher guide there are a multiple-choice and open-ended questions for students along with an answer key. The open-ended questions have many possible answers. The answer key provides only an example of what teachers might expect to find in an answer but does not exhaust all possibilities. Teachers are encouraged to add to or substitute their own questions for these assessments. Finally, each field investigation provides teachers with opportunities for student performance assessment.

Videos

Videos were produced for each chapter of the *Texas Aquatic Science* book. Each video is about two minutes long and is intended to help students visualize the aquatic ecosystems discussed in the book and teacher guide. Many students will not have seen particular ecosystems; thus, video is an effective means to help describe ecosystems and the plants and animals that live in them. In addition, each video graphically illustrates one or two key scientific points made in the corresponding chapter of the book. Links to all videos can be found on www.texasaquaticscience.org, on Apple iTunes U by searching under Texas Aquatic Science, and on YouTube Education by searching under Texas Aquatic Science.

Website

The website, www.texasaquaticscience.org, presents all parts of the book and contains a video portal linking directly to all videos. All illustrations can be viewed and enlarged. The website contains links to partners, and in the tab titled "Teaching Resources" there are links to locations where the teacher guide and supporting materials can be downloaded. Use of materials contained in the website is covered by a copyright and use policy that can be found in the "About" tab.

The Importance of Outdoor Learning

The Texas Natural Resource/Environmental Literacy Plan, launched in 2013 by experts in formal and informal education, encourages learning in outdoor settings. Environmental literacy includes the knowledge, skills, and ability to understand, analyze, and address major natural resource opportunities and challenges. The plan states, "As Texans have fewer direct experiences in the natural world, it becomes increasingly important to educate all citizens about our natural resources through accessible, safe, and enjoyable outdoor experiences. The environment provides an excellent context for this learning."

Whether turning to the research on the effectiveness of learning outdoors or turning to the faces of happy and engaged students and adults, aquatic science educators see the benefits of using the outdoors as a rich learning environment. The outdoors awakens curiosity and discovery. It sharpens focus and concentration. Outdoor projects bring real-world experiences and relevancy to learning. Students of all ages and abilities find connections to the natural world and most easily engage in inquiry-based explorations. Learning outdoors works.

The Texas Aquatic Science curriculum invites teachers to take learning outdoors into natural environments or to bring elements of those environments into the classroom. Curriculum materials include lesson plans for outdoor activities, ideas for class management, and safety suggestions. The involvement of partners TPWD, HRI, and The Meadows Center in education in the outdoors extends throughout the state. The partners offer many venues for experiential learning about aquatic ecosystems, including natural habitat preserves, wildlife management areas, and state parks; glass-bottom boat and kayak tours of Hill Country springs; inland and coastal paddling trails; and discovery voyages into the Gulf. In addition to the tips for finding opportunities for outdoor learning within the curriculum's teacher guide, the project partners and others offer many wonderful training workshops to help teachers overcome barriers to teaching outdoors.

Appendix B

TEKS in Texas Aquatic Science

To support Texas teacher use of the *Texas Aquatic Science* textbook, all materials are aligned with Texas state standards, the Texas Essential Knowledge and Skills (TEKS) for sixth through eighth grades, and for aquatic science and environmental science courses for high school. The TEKS covers topics in biology, chemistry, physics, and other sciences, helping educators meet STEM (science, technology, engineering, and math) teaching requirements. A comprehensive listing of TEKS covered in the teacher guide, by chapter and by lesson, is provided in the guide's introduction. Each lesson in the guide contains its own listing of TEKS covered.

A listing of all TEKS covered in the *Texas Aquatic Science* textbook follows:

6th Grade Science
6.1 A, B; 6.2 A, B, C, D, E; 6.3 A, B, C; 6.4 A, B; 6.5 A; 6.8 B, C, D; 6.9 A, B, C; 6.12 A, C, D, E, F

7th Grade Science
7.1 A, B; 7.2 A, B, C, D, E; 7.3 A, B, C; 7.4 A, B, C; 7.5 A, B, C; 7.6 A; 7.7 C; 7.8 A, B, C; 7.10 A, B, C; 7.11 A, B; 7.12 A, C, D; 7.13 A, B; 7.14 A, B, C

8th Grade Science
8.1 A, B; 8.2 A, B, C, D, E; 8.3 A, B, C; 8.4 A, B; 8.5 D; 8.6 A, C; 8.7 C; 8.9 C; 8.10 A, B; 8.11 A, B, C, D; 8.14 A

Aquatic Science
Aquatic Science: 1 A, B; 2 B, E, F, G, H, I, J, K; 3 A, B, C, D, E; 4 A, B; 5 A, B, C, D; 6 A, B; 7 A, B, C; 8 A; 9 A, C; 10 A, B, C; 11 A, B; 12 A, B, C, D, E

Environmental Science
Environmental Science: 1 A, B; 2 B, E, F, G, H, I, J, K; 3 A, B, C, D, E, G; 4 A, B, C, D, E, F, G, H; 5 A, B, C, E, F; 6 A, C, D, E; 7 A, C, D; 8 A, B, D; 9 A, B, C, D, E, F, G, J, K

Glossary

abiotic. Nonliving; not derived from living organisms; inorganic.
acid rain. Rain or other precipitation containing a high amount of acidity.
acre-foot. A unit of volume used to describe large water resources; an acre-foot is equal to the volume of water it would take to cover an acre of land to a depth of one foot.
adaptation. A behavior or physical trait that evolved by natural selection and increases an organism's ability to survive and reproduce in a specific environment.
aeration. The process of exposing water to air, allowing air and water to mix and water to absorb the gases in air.
aerobic. Occurring or living in the presence of oxygen.
aerobic decomposition. The decomposition of organic matter in the presence of oxygen; decomposition by bacteria that require oxygen.
algae. A group of aquatic organisms ranging from single cell to multicellular that generally possess chlorophyll and are photosynthetic but that lack true roots, stems, and leaves characteristic of terrestrial plants.
algae bloom. A rapid increase in the population and biomass of algae (phytoplankton) in an aquatic system.
anaerobic. Occurring or living in the absence of oxygen.
angler. A person who fishes using a rod, reel, hook, and line.
aphotic zone. The deep part of a lake that does not receive enough light to support photosynthesis.
aquatic ecosystem. A community of organisms together with their physical environment organized around a body of water.
aquatic organism. Any living thing that is part of an ecosystem in water.
aquatic resource. Water and all things that live in or around water.
aquifer. An underground reservoir of water that rests in a layer of sand, gravel, or rock that holds the water in pores or crevices.
arid. Referring to regions of the country characterized by a severe lack of water, which hinders or prevents the growth and development of plant and animal life.
artesian aquifer. A confined aquifer containing groundwater under pressure.
atmosphere. The gaseous envelope surrounding the Earth; the air.
bacteria. A very large group of microorganisms that are called prokaryotic because their cells lack a nucleus and many other organelles, such as mitochondria.
barrier island. A long, narrow island of sand that runs parallel to the mainland.
bay. A body of water partially enclosed by land that is directly open or connected to the ocean.

behavioral trait. The characteristic way an organism acts and reacts in response to its environment, distinguishing one individual or species from another. Behaviors may be inherited or learned; examples are courtship dances in birds and sounds made by many animals.

benthic community, benthos. The community of organisms that live on or in the floor of a body of water, including rivers, lakes, estuaries, and oceans.

benthic macroinvertebrate. An invertebrate visible without the aid of a microscope that lives on or in the bottom substrate.

benthic zone. The bottom substrate of aquatic ecosystems.

biodiversity. The number and variety of living things in an environment.

biofilter. Living material or an organism that captures and biologically degrades pollutants.

biosphere. The part of the world in which life can exist; living organisms and their environment.

biotic. Of or having to do with life or living organisms; organic.

bottomland hardwoods. Wetlands found along rivers and streams of the southeastern and south-central United States where the streams or rivers at least occasionally flood beyond their channels into hardwood forested floodplains.

brackish water. Water that has more salinity than freshwater but not as much as saltwater; often a result of mixing of freshwater and saltwater in estuaries.

buffer. To serve as a protective barrier to reduce or absorb the impact of other influences; something that buffers.

carnivore. An animal that kills and eats other animals.

carrying capacity. An ecosystem's resource limit; the maximum number of individuals in a population that the ecosystem can support.

cartilaginous fish. Fish with a skeleton made of cartilage (a tough, flexible tissue, which is softer than bone) rather than bone; examples are sharks and rays.

cetacean. A marine mammal; examples are whales, dolphins, and porpoises.

channel. The part of a stream where water collects to flow downstream, including the streambed, gravel bars, and stream banks; a dredged passageway within a coastal bay that allows maritime navigation.

channelize. To create an artificial channel through which a stream or river flows using engineered structures to straighten a stream and eliminate its natural tendency to meander.

chemoreceptor. A cell located along the outside surface of a fish that functions as a taste receptor.

chemosynthetic organism. An organism in dark regions of the ocean that gets energy for production and growth from chemical reactions, such as the oxidation of substances like hydrogen sulfide or ammonia, instead of using energy from sunlight and photosynthesis.

chlorophyll. A green pigment found in algae, plants, and other organisms that carry out photosynthesis that enables plants to absorb energy from light.

ciénega. A type of spring-fed wetland that occurs on the desert floor.

Clean Water Act. Primary federal law in the United States governing water pollution, first passed by Congress in 1972.

climate. The weather conditions existing in an area in general, on average, or over a long period of time.

coastal basin. An area that includes parts of coastal plains, peninsulas, and islands that lie adjacent to and between the main river basins for which the coastal basin is named.

cold seep, cold vent. A place where hydrogen sulfide, methane, or other hydrocarbon-rich fluid seeps from the ocean floor.

collector. An aquatic invertebrate that feeds on fine material; examples include caddisfly larvae and mayfly nymphs.

community. A group of plants and animals living and interacting with one another in a particular place.

community fishing lake. In Texas, public lakes 75 acres or smaller located totally within a city, in a public park, or within the boundaries of a state park.

compete. To actively seek and use an environmental resource (such as food) in limited supply that is used by two or more plants or animals or kinds of plants or animals.

condense. To change a gas or vapor to liquid.

confined aquifer. An underground reservoir of water contained within saturated layers of pervious rock material bounded above and below by largely impervious rocks.

conservation, to conserve. The wise use of natural resources such that their use is sustainable long term; includes protection, preservation, management, restoration, and harvest of natural resources; prevents exploitation, pollution, destruction, neglect, and waste of natural resources.

consumer. An organism that feeds on other organisms in a food chain.

contaminant. Impurity, including pollution or pollutants, in a substance such as water; harmful substance in water that can make it unfit for drinking or for supporting aquatic resources.

current. The part of a body of water continuously moving in a certain direction.

dam. A barrier or structure, natural or made by humans or animals, that is placed across a stream or river to prevent water from flowing downstream.

decompose. To decay or rot; to break down or separate into smaller or simpler components.

decomposer. An organism such as a bacterium or fungus that feeds on and breaks down dead plant or animal matter, making essential components available to plants and other organisms in the ecosystem.

decomposition. The process of decaying or rotting; breaking down or separating a substance into smaller or simpler components.

delta, river delta. A low-lying landform created by the deposition of sediment carried by a river where it flows into an ocean, lake, or reservoir; often forms a wetland in freshwater and an estuary in saltwater.

denitrifying bacteria. Microorganisms whose action results in the conversion of nitrates in soil to free atmospheric nitrogen.

detritus. Loose material that results from natural breakdown; material in the early stages of decay.

diatom. A type of algae encased within a cell wall made of silica; most are single-celled, but many can form colonies.

dissolved oxygen. Oxygen gas absorbed by and mixed into water.

dredging. An excavation or digging activity carried out at least partly underwater in shallow-water areas to move bottom materials from one place to another; often used to keep waterways deep enough for boat passage.

drought. An extended period of below-normal rainfall or other deficiency in water supply.

dual doctrine. Two principles or bodies of principles covering the same matter.

ebb tide. The outgoing or receding tide.

echinoderm. Radially symmetrical animal, where the body is like a hub with arms or spines coming out of it; all have rough skin or spines; found only in marine waters; common examples are starfish, sea urchin, and sand dollar.

ecology. Study of the relationships living organisms have with each other and with their environment.

ecosystem. A community of organisms together with their physical environment and the relationships between them.

ecosystem services. Resources and processes that are supplied by ecosystems, generally grouped into four broad categories: provisioning, such as the production of food and water; regulating, such as the control of climate and disease; supporting, such as nutrient cycles and crop pollination; and cultural, such as spiritual and recreational benefits.

ectotherm. An organism that has an internal body temperature at or near the same temperature as the environment in which it lives; internal physiological sources of heat are relatively small and not sufficient to control body temperature.

energy pyramid. A graphic representation that shows the relationship between energy and trophic levels of a given ecosystem.

environmental flow. An amount of freshwater left in a river, estuary, or other water body adequate to support an ecologically sound aquatic environment.

ephemeral stream. A stream that flows, dries up, and flows again at different times of the year.

erosion. The wearing away of land surface materials, especially rocks, sediments, and soils, by the action of water, wind, or ice; usually includes the movement of such materials from their original location.

estuary. A partly enclosed body of water along the coast where one or more streams or rivers enter and mix freshwater with saltwater.

ethical. Following the rules of good conduct governing behavior of an individual or group.

euphotic zone. The upper layer of water in a lake or ocean that is exposed to sufficient sunlight for photosynthesis to occur; also called the photic zone.

euryhaline. An organism that can live in water having a wide range of salinity.

eutrophication. Excessive nutrient input in a body of water that causes excessive plant and algae growth.

evaporation. The process of changing from a liquid state into vapor.

exoskeleton. The external skeleton or covering that supports and protects an animal's body.

extinction. The end of a species.

fault. A fracture or crack in rock in the land surface, across which there has been significant displacement along the fracture as a result of Earth movement.

filter feeder. An aquatic animal, such as an oyster and some species of fish, that feed by filtering tiny organisms or fine particles of organic matter from water that passes through the animal.

fin. A wing- or paddlelike part of a fish used for propelling, steering, or balancing in the water.

first-order stream. A small stream with no tributaries coming into it.

flood tide. The incoming or rising tide.

floodplain. The flat land on both sides of a stream onto which the stream's extra water spreads during a flood.

food chain. A series of plants and animals linked by their feeding relationships.

food web. Many interconnected food chains within an ecological community.

freshwater. Water with a salt content lower than about 0.05%; for comparison, seawater has a salt content of about 3.5%.

freshwater inflow. Less saline water that flows downstream in streams and rivers, or as runoff, and enters estuaries and bays.

fry. Newly hatched fish.

fungi. Members of a large group of organisms that include microorganisms such as yeasts and molds, as well as mushrooms; classified separately from plants, animals, and bacteria.

geosphere. The solid part of the Earth consisting of the crust and outer mantle.

gill. A respiratory organ that enables aquatic animals to take oxygen from water and to excrete carbon dioxide.

grazer. An aquatic invertebrate such as a snail or water penny that eats aquatic plants, especially algae growing on surfaces.

groundwater. Water that flows or collects beneath the Earth's surface in saturated soil or aquifers.

habitat. The natural environment in which an organism normally lives, including the surroundings and other physical conditions needed to sustain it.

halophyte. A plant that grows in waters of high salinity.

headwaters. The high ground where precipitation first collects and flows downhill in tiny trickles too small to create a permanent channel; the place where spring water flows from an aquifer and starts streams.

herbivore. An animal that eats plants.

humidity. The amount of water vapor in the air.

hydrologic cycle. The natural process of evaporation and condensation, driven by solar energy and gravity, that distributes the Earth's water as it evaporates from bodies of water, condenses, precipitates, and returns to those bodies of water.

hydrosphere. All of the Earth's water, including surface water, groundwater, and water vapor.

hypersaline. Having salt levels surpassing that of normal ocean water (more than 3.5% salts).

hypoxia. The condition in water where dissolved oxygen is less than 2 to 3 milligrams per liter.

hypoxic zone. An area in which the water contains low or no dissolved oxygen, causing a condition known as hypoxia.

impervious. Not permitting penetration or passage; impenetrable.

impoundment, to impound. A reservoir created in a river valley by placement of a dam across the river.

indicator species. A species that defines a trait or characteristic of the environment, including an environmental condition such as pollution or a disease outbreak.

inorganic. Composed of matter that does not come from plants or animals either dead or alive; abiotic.

intermittent stream. A stream that flows, dries up, and flows again at different times of the year.

invasive species. A species that has been introduced by human action to a location where it did not previously occur naturally, has become capable of establishing a breeding population in the new location without further intervention by humans, and has spread widely throughout the new location and competes with native species.

invertebrate. Any animal without a spinal column; examples are insects, worms, mollusks, and crustaceans.

irrigation. The application of water to the land or soil to assist in the growing of agricultural crops, watering of lawns, and promoting plant growth in dry areas and during periods of inadequate rainfall.

karst aquifer. An underground reservoir of water contained in limestone and marble rocks that are filled with numerous small channels and, in some cases, large underground caverns and streams.

lagoon. A body of saltwater separated from the ocean by a coral reef, sandbar, or barrier island.

lake. A large body of standing water.

larva. The newly hatched form of many fish, insects, or other organisms that have a distinct separate life stage before metamorphosis, or change, into adults; wingless, often wormlike form of many insects before metamorphosis that have no wings and cannot reproduce.

lateral line. An organ running lengthwise down the sides of fish, used for detecting vibrations and pressure changes.

lentic water. Referring to water that is not flowing, such as a pond or lake.

limestone. A sedimentary rock composed largely of calcium carbonate; often composed from skeletal fragments of marine organisms such as coral; slightly soluble in water and weak acid solutions, which leads to karst landscapes, in which water erodes the limestone over millions of years and creates underground cave systems.

limnetic zone. The part of a lake that is too deep to support rooted aquatic plants.

littoral zone. The part of a lake that is shallow enough to support rooted aquatic plants.

macroinvertebrate. An invertebrate large enough to be seen without the use of a microscope.

mainland. A large landmass located near smaller landmasses such as islands.

marble. A hard, crystalline, metamorphic form of limestone rock.

marsh. A wetland dominated by reeds and other grasslike plants.

metamorphosis. The process of transformation from an immature form to an adult form in two or more distinct stages. For insects, complete metamorphosis results in little resemblance between the larva and adult; in incomplete metamorphosis, the larva resembles the adult.

microorganism. A very tiny organism, such as a one-celled bacterium and fungus, that can be seen only by using a microscope.

migratory, migration, to migrate. Referring to an animal, fish, or other organism that undertakes a regular seasonal journey, such as are made by many species of ducks.

milt. The seminal fluid containing sperm of male fish and aquatic mollusks that reproduce by releasing this fluid onto nests or water containing eggs.

mollusk. A member of a large group of invertebrate animals that includes hard-shelled animals such as clams, oysters, scallops, mussels, and snails; soft-bodied animals such as octopus that find shelter in cavities; and soft-bodied animals such as squid that are free-swimming.

mussel. A mollusk that attaches to objects or to other mollusks, often in dense clusters, and has two shells that close on each other; similar to a clam.

natural physiographic region. A region based upon physical geography and natural features of terrain and habitats.

natural resource. Something that is found in nature that is useful to humans.

natural selection. The natural process in which those organisms best adapted to the conditions under which they live survive and poorly adapted forms are eliminated.

neotropical migratory bird. A bird that breeds in Canada or the United States during the summer and spends the winter in Mexico, Central America, South America, or the Caribbean islands.

niche. The function, position, or role of a species within an ecosystem.

nitrate. A salt of nitric acid; produced for use as fertilizers in agriculture. The main nitrates are ammonium, sodium, potassium, and calcium salts.

nonindigenous species. An organism that has been introduced to an area to which it is not native; an exotic or nonnative species.

nonpoint source pollution. Water pollution that comes from a combination of many sources rather than a single outlet.

nutrient. A chemical, organic, or inorganic compound, that an organism needs to live and grow that is taken from the environment.

nymph. The juvenile stage of development of certain insects; a nymph looks generally like the adult except it is smaller, has no wings, and cannot reproduce.

ocean basin. Large depression below sea level containing saltwater.

omnivore. An animal that eats both plants and animals.

organic matter. Material that comes from plants or animals either dead or alive that is capable of decay; important in the transfer of nutrients from land to water.

osmoregulation. A physiological adaptation of many organisms that allows them to regulate their intake of salts or freshwater to keep their fluids, such as blood, from becoming too salty or too dilute.

oxbow lake. Crescent-shaped lake formed when a bend of a stream is cut off from the main channel.

parasite. An organism that lives on or in the living body of another species, known as the host, from which it obtains nutrients.

pelagic fish. A fish that lives near the surface or in the water column that almost constantly moves about; not associated with the bottom.

perennial stream. A stream that flows for most or all of the year.

periphyton. A complex mixture of algae, detritus, bacteria, and microbes that are attached to submerged objects in most aquatic ecosystems.

permeable. Having pores or openings that permit liquids or gases to pass through.

pervious. Allowing water to pass through.

photic zone. The upper part of a lake where enough light penetrates the water to allow photosynthesis to occur.

photosynthesis. A process used by plants, algae, and many species of bacteria to convert energy captured from the sun into chemical energy that can be used to fuel the organism's activities; photosynthesis uses carbon dioxide and water, releasing oxygen as a waste product; sugars or carbohydrates are a by-product of photosynthesis.

phytoplankton. Algae and plant plankton, including single-celled protozoa and bacteria.

plankton. Microscopic, free-floating, living organisms not attached to the bottom or able to swim effectively against most currents.

playa lake. A round hollow in the ground in the Southern High Plains of the United States that fills with water when it rains, forming a shallow, temporary lake or wetland.

point source pollution. Water pollution that comes from a single source or outlet.

pollutant. A substance that contaminates the water, air, or soil.

pollution, to pollute. The contamination of air, water, or soil by substances that are harmful to living organisms, especially environmental contamination with human-made waste or chemicals; also the harmful substances themselves.

polychaete. A marine worm in which each body segment has a pair of protrusions called parapodia that bear bristles made of chitin; sometimes called a bristle worm.

pond. A body of standing water small enough that sunlight can reach the bottom across the entire diameter.

pond succession. The natural process by which sediment and organic material gradually replace the water volume of a pond, ultimately resulting in the area becoming dry land.

pool. Part of a stream where the water slows down, often with water deeper than the surrounding areas, which is usable by fish for resting and cover.

population. A group of individuals of the same species occupying a specific area.

porous. Full of pores, holes, or openings that allow the passage of liquid or gas.

precipitation. A form of water such as rain, snow, or sleet that condenses from the atmosphere and falls to Earth's surface.

predator. An animal that lives by capturing and eating other animals.

prey. An animal that is eaten by a predator.

primary consumer. An animal that eats plants; an herbivore.

producer, primary producer. An organism that is able to produce its own food from nonliving materials and that serves as a food source for other organisms in a food chain; examples are green plants, algae, and chemosynthetic organisms.

protozoa. A group of microscopic single-celled organisms that are called eukaryotic because their cells are organized into complex structures by internal membranes, the most characteristic of which is the nucleus.

rapid. An area of very turbulent flow; a part of a stream where the current is moving at much higher velocities than in surrounding areas and the surface water is greatly disturbed by obstructions that reach above the surface, or nearly so.

recharge. Water that soaks into and refills an aquifer.

resaca. A former channel of the Rio Grande that has been cut off from the river and has filled in with silt and water to create shallow marshes and ponds.

reservoir. An artificial or natural lake built by placing a dam across a stream or river and used to store and often regulate discharge of water; underground storage area of water, such as in an aquifer.

riffle. A relatively shallow part of a stream in which the water flows faster and the water surface is broken into waves by obstructions that are completely or partially underwater.

riparian vegetation. The plant community next to a stream, starting at the water's edge and extending up the bank and beyond on either side of the stream.

riparian zone. Land next to a stream, starting at the top of the bank and containing vegetation on either side.

river. A large stream.

river basin. A drainage area, generally made up of many smaller units called watersheds; area of land drained to form a river.

rule of capture. A landowner's legal right to pump and capture groundwater or runoff before it enters a stream.

run. A portion of a stream that has a fairly uniform flow and generally smooth surface water, with the slope of the water surface generally parallel to the overall slope of the section of stream.

runoff. Precipitation, snowmelt, or other water that flows onto the land but is not absorbed into the soil.

saltwater. Water with about 3.5% salt content, such as ocean water or seawater.

sand sheet wetlands. Small isolated depressions found in places where wind erodes away topsoil, exposing clay soils underneath that trap and hold water when it rains.

saturated. Soaked with moisture; having no pores or spaces not filled with water.

scale. Any of the small, stiff, flat plates that form the outer body covering of most fish.

scavenger. An animal that eats the organic material of dead plants and animals.

scraper. An aquatic invertebrate that has special mouthparts used to remove algae or other food material growing on the surface of plants or solid objects; the mouthparts act like a sharp scraper blade.

seafood. Any sea life used by humans as food; includes fish and shellfish.

seagrass. Submerged rooted aquatic plants that tolerate salinity.

sediment. Silt, sand, rocks, and other matter carried and deposited by moving water.

sedimentation. The process of particles carried in water falling out of suspension; deposition of silt, sand, rock, and other matter carried by water.

seep. A place where water oozes from springs in the ground.

shellfish. Aquatic invertebrates that have exoskeletons and are used as food, including various species of mollusks and crustaceans, such as crabs, shrimp, clams, and oysters.

shipping lane. A regular route used by oceangoing ships.

shredder. An aquatic invertebrate such as a stonefly nymph that feeds by cutting and tearing organic matter.

silt. Tiny specks of dirt, sized between sand and clay particles, that can be suspended in water or fall out of suspension to cover plants and the bottom of lakes or pool sections of rivers and streams.

slough. A backwater or secondary channel of a stream.

spawn. To release eggs and sperm, usually into water by aquatic animals, including fish.

species. A group of related individuals sharing common characteristics or qualities that interbreed and produce fertile offspring having the same common characteristics and qualities as the parents.

sport fisherman. An angler who catches fish for personal use or recreation rather than to make a living.

spring. A place where groundwater flows to the surface of the Earth or where an aquifer meets the ground surface.

stakeholder. An individual, group of people, or organization that has an interest in, has a concern about, or will be affected by an action or issue.

stock, stocking. To introduce fish that have been produced elsewhere to a body of water; to add a new species to a water body or increase the number of individuals of a fish species already present in a water body.

storm surge. A rise in the height of ocean water associated with high storm winds pushing against the ocean water; flooding caused by high ocean waters in coastal areas.

stream. A body of flowing water.

stream bank. The shoulderlike sides of a stream channel from the water's edge to the higher ground nearby.

stream flow. Water flow in a stream.

streambed. The bottom of a stream or river channel.

structural trait. One of the internal and external physical features that make up an organism, including shape, body covering and internal organs, which can determine how an organism interacts with its environment.

subsidence. A lowering, compaction, or collapse of the ground surface caused when large amounts of groundwater are withdrawn from an aquifer.

surface water. Precipitation that runs off the land surface and is collected in ponds, lakes, streams, rivers, and wetlands.

swamp. A wetland in which trees or woody shrubs predominate.

swim bladder. An air-filled sac near the spinal column in many fish species that helps maintain the fish's position in the water column.

tide, tidal. The rise and fall of sea levels caused by the rotation of the Earth and the gravitational forces exerted by the Earth, moon, and sun on the ocean.

toxic algae bloom. An algal bloom that has a negative impact on other organisms due to production of natural toxins by some species of algae.

transpiration. The passage of water through a plant to the atmosphere.

tributary. A stream that flows into a larger stream or other body of water.

trophic level. A group of organisms that occupy the same position in a food chain; each step of an energy pyramid.

tunicate. A marine filter-feeding organism that has a saclike body; most species attach to rocks or other hard structures on the ocean floor, while some are pelagic (free-swimming); commonly called sea squirt.

turbid. Referring to water that is so full of small particles, such as silt, that the water is no longer transparent but appears cloudy.

turbulent water flow. Water flow where water at any point is moving in many different directions at once and at differing velocities.

unconfined aquifer. An underground reservoir of water that is directly connected to the surface and has water levels dependent on relatively constant recharge.

unicellular. Referring to an organism that consists of just a single cell; this includes most life on Earth, with bacteria being the most abundant.

vertebrate. An animal that has a backbone or spinal column.

wastewater treatment facility. A place that treats wastewater from homes and businesses, such as toilet or sewage water.

water pollution. An excess of natural or human-made substances in a body of water; especially, the contamination of water by substances that are harmful to living things.

water quality. The fitness of a water source for a given use, such as drinking, fishing, or swimming.

water table. The surface of the subsurface materials that are saturated with groundwater in a given vicinity.

waterfowl. Ducks, geese, and swans.

watershed. All the land from which water drains into a specific body of water.

watershed address. The watershed, sub-watershed, and sub-sub-watershed that includes a particular location.

weather. The condition of the atmosphere at any point in time, such as if it is hot or cold, raining or dry, windy or calm, or clear or cloudy.

wetland. A low-lying area where the soil is saturated with water at least seasonally and that supports plants adapted to saturated soils.

zooplankton. Animal plankton, including single-celled and complex multicellular organisms.

Index

abiotic, 58
acid rain, 13
acre-foot, 163
adaptations: of aquatic plants, 35, 97, 106–107; of birds, 135; to dark underground aquatic ecosystems, 73; defined, 28; of fish, 29–32, 43, 47, 54, 74, 85, 97, 107, 121, 123; of oysters, 118; of sea turtles, 135; stream plants and animals, 85; through natural selection, 54–55; of turtles, 107; wetland animals and plants, 107–109
aeration, 8
aerobic, 103
aerobic bacteria, 103
aerobic decomposition, 95
agriculture: irrigation water sources, 61, 69, 79, 164; irrigation water use historically, 159; pollution from, 9, 71, 131, 158; water return through transpiration, 12; water used for food production/ranching, 5–6, 164
algae, 35, 54, 87, 88
algae blooms, 121, 131, 135
alligators, 41, 88
Amazon River, 86
American alligator, 41
anaerobic, 103
anaerobic bacteria, 103
anaerobic decomposition, 99
angler, 145
Angler Recognition Program (TPWD), 152–53
animal husbandry, water used for, 6, 164
animals, water conservation methods, 16, 17
aphotic zone, 95

aquatic insects, 33–34, 86–89, 98
aquatic invertebrates, 32–35
aquatic organisms, 28, 46–47, 53
aquatic plants: adaptations, 35, 97, 106–107; endangered, 86; in ponds and lakes, 94–98; specialist and generalist, 47; in streams, 86, 87; survival needs, 97. *See also specific plants*
aquatic resources, 7, 153
aquatic science careers: aquatic science laboratory technician, 57; benthic ecologist, 128; drinking water treatment plant workers, 20; educator at an aquatic science nature center, 80; endangered species protection worker, 66; environmental protection worker, 66; fish and wildlife conservation officer (game warden), 156; fisheries biologist, 45; fish hatchery technician and biologist, 101; hydrologist, 26; marine biologist, 144; park ranger, 166; stream ecologist, 91; wastewater treatment plant workers, 20; water quality regulator, 10; wildlife technician and biologist, 114
aquatic science laboratory technician, 57
aquatic species protection agencies, 160
aquifer ecosystems, 59, 71, 73–75
aquifer recharge, 19, 60, 68, 70, 78–79
aquifers: artesian, 70; carrying capacity, 73; confined, 70; defined, 60, 67; depletion, 75, 77–78, 162; functions of, 67–68; groundwater and, 67–70; importance of, 69; karst, 71, 73–74; major and

aquifers (cont.)
  minor, 68–69; photosynthesis in, 71, 73–74; Texas, 6, 59, 60, 68–71; unconfined, 70; water purification by, 79; wetlands and, 111
Aransas Bay, 116, 117, 120
Aransas National Wildlife Refuge, 53
arid, 7
Army Corps of Engineers, 160
arrowheads, 106
artesian aquifers, 70
artificial reefs, 136–37
assessment, Texas Aquatic Science curriculum, 171
atmosphere, 11, 16
Austin, 76
Austin blind salamander, 74

bacteria, 71
Balcones fault, 71
Balmorhea spring-fed swimming pool, 75
Balmorhea springs, 70
barrier islands, 63, 117, 121, 123, 125, 132
Barton Springs, 76
Barton Springs blind salamander, 74
Barton Springs Pool, 75, 76
bay ecosystems, 59, 63, 115–16
bay habitats, 118–19, 125, 127
bays: coastal, 116–17; defined, 63, 115; environmental threats, 63, 64; fish species in, 123; freshwater inflow, 117–18, 140; importance of, 125–26; miles of, 145; pollution in, 121; salinity, 116; Texas, 22, 23; tides in, 117–18; wind on, 126
beaches, 132
beach trash, 142–43, 166
beavers, 103
behavioral trait, 28
benthic community, 86
benthic ecologist, 128

benthic macroinvertebrates, 86, 88–89
benthos, 86
biodiversity, healthy ecosystems and, 64–65
biofilters, 121
biologists: fisheries, 45; fish hatchery, 101; marine, 144; wildlife, 114
biosphere, 12
biotic, 58
birds: adaptations, 135; migratory, 109–10, 123–25
black fly larvae, 88–89
black-necked stilt, 107, 109
bloodworms, 88–89
blue catfish, 39
blue crab, 53, 119, 123
bluegills, 50, 54
Blue Hole, 72
blue-winged teal, 110, 111
boiling point, 3
bony fish, 32
bottlenose dolphins, 44
bottomland hardwoods, 105
Bracketville, 77
brackish water, 35, 63
Bryde's whale, 135
buffer, 83

caddisfly, 89
Caddo Lake, 55, 93–94
carnivores, 50
carrying capacity, 49, 73
cartilaginous, 32
catch and release, 151, 153, 154
catfish: adaptations, 30, 47, 54, 74, 85; blue, 39; breeding methods, 48; channel, 54; feeding behaviors, 39, 88, 97; fishing for, 146, 150; flathead, 48, 49; niche environment, 47–49, 53; senses, 39; species variety, 88; stocking, 153; toothless blindcat, 74
cattails, 106, 108
caudal fin, 30, 31

Central Flyway, 109–10
Central Texas wetlands, 105
cetaceans, 138
channel catfish, 54
channelizing, 111
chemoreceptors, 73
chemosynthetic organisms, 133
chlorophyll, 36
ciénegas, 105
citizen conservation groups, 110, 122, 158
Clean Water Act, 10, 26, 158
climate, Texas, 14
clouds, 3, 12
coastal basins, 22–23
coastal bays, 116–17
coastal wetlands: biodiversity illustrated, 103; formation of, 104, 106; geographic range, 116; loss of, 140; migratory bird habitats, 109–11, 124–25; shoreline erosion, 63, 64
coastline, miles of, 145
cohesion, 3
cold seeps, 133
collectors, 87
Comal Springs, 76
Comanche Springs, 162
Comanche Springs pupfish, 70
commercial fisheries, 140
community, 47–49
community fishing lakes, 153
compete, 47
competition, 48–49, 53–56
condenses, 12
confined aquifers, 70
conservation: by animals, 16, 17; of aquatic resources, 153; citizen groups, 110, 122, 153, 158; defined, 7; funding, 151; groundwater conservation districts, 160, 162; opportunities for individuals, 166; oyster shell recycling, 127, 166; planning for future water needs, 164–66; rainwater, 17; wetlands, 114

conservation officer, 156
consumers, 50
contaminants, 5
coral reefs, 135–37
Corpus Christi Bay, 116, 117, 120
crab: blue, 53, 119, 123; economic importance of, 121; in the food chain, 51–53, 123, 138; protective habitats, 35, 118–19, 121; recreational catches, 119; sand, 132
crickets, 147
crustaceans, 34
currents, 135, 142–43
cypress, 107
Cypress Creek, 72, 74

dams, 75, 89, 92–93, 163
damselfly, 88
darters, 74, 85, 87
dead zone, 63, 132
decompose, 9
decomposers, 50–52, 54, 96, 103, 131
decomposition, 95, 103
Del Rio, 77
deltas, 115
denitrifying bacteria, 71
detritus, 50, 54, 95, 103
diatoms, 36
diffused surface water, 163
dinoflagellates, 36
dissolved oxygen, 95
dolphins, 44, 138
dorsal fin, 30, 31
dragonfly, 88
dredging, 117
drinking water, 19
drinking water treatment plant workers, 20
drought, 7, 14–15, 112
dry streambed, 23
dual doctrine, 159–61
ducks, 107, 124
Duck Stamps, 166
Ducks Unlimited, 114
duckweed, 106

ears, fish, 32
earth: amount of water on the, 4–6; surface collapses, 18; time of fish on, 32
earthquakes, 71
Eastern oyster, 41, 119
East Matagorda Bay, 120
East Texas wetlands, 105
ebb tide, 117
echinoderms, 136
ecologists: benthic, 128; stream, 91
ecology, 58
economy: coral reefs and the, 135–37; fishing and fisheries importance to, 63, 64, 93–94, 100, 125, 140; historical role of springs in the, 76–77; marine, 125–27, 140; oil and gas, 125, 136; recreation and the, 61, 132–33; rivers and streams importance to the, 89–90; shellfish and the, 64, 121, 140; tourism and the, 64, 100, 110, 112, 125
ecosystems, 58
ecosystems, aquatic: abiotic and biotic parts of, 58; biodiversity in, 64–65; biodiversity's role in healthy, 64–65; energy flow through, 50–52; Mississippi River's influence on, 130; niche environment, 47–48; rainfall's influence on, 82; resource limits, 49; Texas, 59; threats to, 55–56. *See also specific types, e.g.* rivers
ecosystem services, 125
ectotherms, 29
educators at aquatic science nature centers, 80
Edwards Aquifer, 38, 70, 71, 72, 74, 79
electricity/electrical power, 6
El Niño, 15
El Paso, inches of rain per year, 7
emergent plants, 97
endangered species: adaptations, 47; in aquifers, 79; fish, 70; protecting, 66, 138–39, 156; sea turtles, 43, 138–39; Texas' ranking for number of aquatic, 165; whooping crane, 53
endangered species protection worker, 66
energy pyramid, 50–52
environmental flow, 163–64
environmental literacy, 172
environmental protection worker, 66
Environment Protection Agency (EPA), 10, 160
ephemeral (intermittent) streams, 23, 82
erosion, 4, 25, 63–64, 83, 98–99
estuaries: defined, 63, 115; draining into the Gulf, 130; environmental threats, 63, 64; fish species in, 121, 123; freshwater inflow, 117–18, 140; importance of, 125–26; miles of, 145; pollution in, 121; tides in, 117–18; wind on, 126
estuary ecosystems, 59, 63, 115–16
estuary habitats, 118–19, 123–25, 127
ethical, 151
ethical anglers, 151–52
euphotic zone, 36, 95
euryhaline, 29
eutrophication, 98
evaporation, 12
exoskeleton, 34
extinction, 79

faults, 71
Federal Water Pollution and Control Act, 158
feeding specialization, 53
field investigations, Texas Aquatic Science curriculum, 171
fifth-order streams, 88
filter feeders, 97, 135
fins, 30, 31, 85
first-in-time, first-in-right rule, 159–61

first-order streams, 85–88
fish and wildlife conservation officer (game warden), 156
Fish and Wildlife Service, US, 160
fish bait, 147
fish basics: buoyancy, 30; determining age of, 31; fins, 30, 31, 85; internal organs of, 28–29; sensory abilities, 32, 39; skin, 31; swimming ability, 30; time on earth, 32
fish conservation, funding, 151
fisheries biologist, 45
fishermen: ethical anglers, 151–52; sport fisherman, 145
fish hatchery technician and biologist, 101
fishing: catch and release, 151, 153, 154; common recreational catches, 119; daytime vs. nighttime, 150; economic importance of commercial and recreational, 63, 64, 93–94, 100, 125, 140; effect of light on, 150; ethical, 151–52; finding the fish when, 148–50; fun when, 145–46; rules when, 151–52; saltwater, 140, 147, 148–50; success at, 146–48; weather and, 150; wetlands, 112
fishing licenses, 151, 166
fish species: adaptations made by, 29–32, 43, 47, 54, 74, 85, 97, 107, 121, 123; common in streams, 85; endangered, 70; environmental contaminants in, 155; filter feeders, 97, 135; Gulf of Mexico, 140; niche communities, 47–48; number of, 32, 84; pelagic, 135; percent at risk, 165; population surveys in lakes, 100. *See also specific species*
flathead catfish, 48, 49
flood control, 111, 114
flooding, 14, 111
floodplain, 83, 84, 105

flood tide, 117
Flower Garden Banks National Marine Sanctuary, 135–37
food chain, 36, 50–52, 96–97, 131
food web, 50–51, 86–88, 98
forested wetlands, 105
Forest Service, Texas and US, 160
Fort Clark, 77
fountain darters, 74
fourth-order streams, 88
freezing point, 3
freshwater, 5
freshwater fish: adaptations, 29, 32, 43; dependence on wetlands, 104, 109; fishing for, 147; number of, 32; at risk statistics, 165. *See also specific species*
freshwater inflow, 63, 117–18, 123, 130, 140
freshwater insects, 33
freshwater-saltwater mixing, 63, 117–18
freshwater wetlands, 104
frogs, 107, 109, 147–48

Galveston Bay, 116, 117, 120, 121
game warden, 156
gas (water vapor), 2
geosphere, 18
giant salvinia, 55
gills: as adaptation, 38, 75, 97; aquatic insects, 33; function, 29–30, 41, 121; handling in catch and release, 154; in oysters, 41, 121; in sea turtles, 43
gizzard shad, 47–48, 97
grasshoppers, 147
grazers, 87, 88, 97, 98
green sea turtle, 138
groundwater: aquifers and, 69–70; defined, 18; depletion, 18–19; impact of withdrawals, 75; legal right to own and use, 159, 162; sources of pollution in, 79; Texan's dependence on, 19; wetlands and, 67–68

groundwater conservation districts, 160, 162
groundwater management, 160
groundwater recharge, 19, 78, 111
Guadalupe bass, 39
Guadalupe River, 38
Gulf Coast: beaches, 132, 142; seashore, 132–33; wetlands, 106
Gulf Intracoastal Waterway, 125, 126
Gulf of Mexico: aquatic life, 132; biodiversity, 63–64, 133; circulation patterns influence on Port Mansfield Channel, 118; coral reefs, 135–38; currents, 135; depth, size, and shape, 129–30; ecological threats and resilience, 140–41; economic importance of, 132–33, 140; fish species, 140; freshwater inflow, 130; habitat creation, 136–37; importance of, 130; marine mammals, 138; recreational importance of, 132–33; rivers inflow into, 22; sea turtles, 138–40; surrounding states and islands, 129–30
Gulf of Mexico ecosystem, 59, 63–64, 130

habitats: bays and estuaries, 118–19, 123–25, 127; creating new, 136–37; defined, 47; migratory birds, 109–11, 124–25; protective, 35, 118–19, 121; seagrass, 121; sea turtles, 132, 136; wetlands, 118–19, 123
halophytes, 106
Harte Research Institute for Gulf of Mexico Studies, Texas A&M University-Corpus Christi, xii, 127
headwaters, 23
headwater streams, 86, 87
health risks: environmental contaminants in fish, 155; oysters, 121; toxic algae blooms, 135

hearing, fish, 32
herbivores, 50
higher-order streams, 88
High Plains wetlands, 105
human body: amount needed to live per day, 5; risks from seafood consumption, 121, 155; survival needs, 46; toxic algae blooms and the, 135; water in the, 2, 4
humidity, 15
hunters/hunting, 110, 114
hurricanes, 14
hydrologic cycle, 4, 11–12
hydrologist, 26
hydrosphere, 4
hypersaline, 116
hypoxia, 132
hypoxic zone, 63, 131, 132, 140–41

ice, properties of, 2, 3–4
impervious/impervious rock, 70
impounded, 61
inorganic, 9
insects: aquatic, 33–34, 86–89, 98; freshwater, 33; omnivorous, 98; predator, 98
instream flow, 163
intermittent (ephemeral) streams, 23, 82
invasive species, 55–56, 75, 160
invertebrates, 32–35, 97, 136
irrigation: agricultural, 5–6, 61, 69, 79, 159, 164; aquifers for, 69, 79; defined, 61; early water systems, 159; lakes and ponds for, 61; reductions in use through conservation, 164; water used for, 5

Jacob's Well, 72, 74, 162
jawless fish, 32
jellyfish, 142–43

karst aquifers, 71, 73–74
Kemp's ridley sea turtle, 43, 125, 138–40, 166

lagoons, 116, 123
Laguna Madre, 116, 118, 120, 121
lake ecosystems, 59, 61–62, 96–97
Lake Fork, 100
lakes: all about, 95–96; aquatic plants, 95–98; community fishing, 153; creation of, 92, 93, 94; defined, 61, 95; fish population surveys in, 100; functions of, 93; impounded, 61; maintaining the health of, 99; number of, 145
Lake Texoma, 56
lake zones, 95–96, 97–98
largemouth bass: breeding methods, 48; feeding behaviors, 40, 47–48; fishing for, 40, 100, 146, 147–48; in the food chain, 50, 52, 54, 97; habitat, 40; niche environment, 47–48; swimming power, 30
larvae, 87
Las Moras Springs, 77
lateral line, 32
lentic water, 92
Leona Springs, 77
light zones, lakes, 95–96
limestone, 70, 71, 74
limnetic zone, 95
lionfish, 55, 136
littoral zone, 95, 97–98
loggerhead sea turtle, 138
Lower Colorado River Authority, 160

mackerel, 135
macroinvertebrates, 86, 88–89, 121
mainland, 117
mallards, 42
mangrove, 106
manta rays, 136
marble, 70, 71, 74
marine biologist, 144
marine mammals, 138, 166
marshes, 104
Matagorda Bay, 116, 117, 120
mayfly, 88, 89
Meadows Center for Water and the Environment, Texas State University, xii
menhaden, 135
metamorphosis, 33
Mexican tetra, 70
microalgae, 36
microorganisms, 71
migratory, 109
migratory birds, 109–10, 123–25
milt, 48
minnows, 87, 147
Mississippi Flyway, 109–10
Mississippi River, 63, 86, 140–41
Mississippi River delta, 115
Mississippi River watershed, 86, 130–31
mollusks, 33, 88, 96–97, 136
mosquitofish, 107
mussels, 56, 85, 88–89, 97

National Oceanic and Atmospheric Administration (NOAA), 15
National Resource Conservation Service, 160
natural physiographic regions, 25–27
natural resource, 6
natural resource management, 62–64
natural selection, 54–55
natural surface water, 161–62
New Braunfels, 76
niche environment, 47–49
nitrates, 9
nonindigenous species, 55
nonpoint source pollution, 17, 24, 83–84, 89, 158
nonpoint source pollution controls, 158–59
Nueces Bay, 116
nutrients, 61, 74–75
nymph, 34, 87–88

ocean basin, 129
ocean ecosystems, 59, 63–64
oceans, percent of rain falling on the, 13

Ogallala Aquifer, 69, 70, 79, 105
oil and gas production, 125, 130, 135–37, 140, 141
oil spills, 141
omnivores, 50
omnivorous insects, 98
orange cup coral, 136
orangethroat darter, 38
organic matter, 4, 8–9
osmoregulation, 29
Outdoor Annual (TPWD), 151
oxbow lakes, 105, 106
oxygen, dissolved, 95
oyster reefs: creation of, 118, 121; formation of, 41; functions of, 41, 121, 123; loss of, 63, 64, 125, 140; red drum and, 43
oysters: adaptations, 118; Eastern, 41, 119; economic importance of, 64, 121, 140; feeding behaviors, 41; recreational catches, 119; unsafe to eat, 121
oyster shell recycling, 127, 166

paddlefish, 97
Padre Island, 118, 121, 132, 138, 142
parasites, 50
park ranger, 166
Pecos gambusia, 70
pectoral fins, 30, 31, 85
pelagic fish, 135
pelvic fins, 30, 31
perennial stream, 23
periphyton, 121
permeable, 67
pervious/pervious rock, 70
petroleum industry, 125, 130, 135–37, 140, 141
photic zone, 95, 96
photosynthesis, 36–37, 50, 71, 73–74, 95–96, 121
phytoplankton, 35, 36–37, 95–97, 130–32, 135
plankton, 34, 96–97, 118, 119
plants, use of rain by, 16

playa lakes, 70
playa lake wetlands, 105
point source pollution, 24, 89
pollutants: beach trash, 142–43; defined, 158
pollution (water): causes of, 4; filtering out, 18, 24, 79, 111, 121; freshwater, 5; legislation addressing, 10, 26, 158; monitoring, 158; rivers and streams, 89
pollution (water) sources: agriculture, 9, 71, 98–99, 131, 158; heat, 8–9; individuals, 98, 130, 131, 158; nonpoint source, 17, 24, 83–84, 89, 158; nutrient load, 140–41; oil and gas production, 136; organic matter, 8–9; point source, 24, 89; rainfall, 13; sediment, 24–25
polychaetes, 136
pond ecosystems, 59, 61–62, 96–97
ponds, 61, 92–99
pond succession, 98–99
pools, 75, 76, 82, 83
populations: defined, 47; resource limits, 49, 52
porosity, 78
porous, 17
Port Arthur, inches of rain per year, 7
Port Mansfield Channel, 118
prairie wetlands, 109
precipitation, 13. *See also* rain
predation, 53–54
predator, 54
predator insects, 98
prey, 54
primary consumers, 50
primary producers, 36
prior appropriation, 159, 161
producers, 36, 50
protozoa, 71

rain: ecosystems influence, 82; fishing in the, 150; individuals conservation of, 17; percent as

groundwater, 18; percent falling on the land, 13; percent returned the atmosphere, 16; pollution and, 12–13; stream size and flow influenced by, 82; in Texas, 7, 14, 82; toxic, 13
raindrops, 12, 25
ranchers, water used by, 6, 164
rapids, 82
recharge areas, 19, 60, 68, 70, 78–79, 111
recreation: bays and estuaries, 126; economic importance of, 61, 89, 93, 132–33; Gulf of Mexico, 132–33; lake-related, 93, 100
recycling: the hydrologic cycle, 4, 11–12; oyster shell recycling, 127; by scavengers and decomposers, 50
reddish egrets, 123
red drum/redfish, 29, 43, 118–19, 121, 123, 148, 151
Redfish Bay, 120
redhead duck, 123
Red River, 22, 81–82, 93–94
Red River Groundwater Conservation District, 160
red tides, 135
reef invertebrates, 136
resacas, 105
reservoir management, 160
reservoirs, 61, 89, 92–93
riffles, 82, 83
Rio Grande, 22, 82, 105
riparian law, 159–61, 163
riparian vegetation, 83–84
riparian wetlands, 105
riparian zone, 83–84
river basins, 22–23, 26, 81, 130
river cooter, 40
river ecosystems, 59, 60–61, 84–85
river otters, 85
rivers: as borders, 22, 81–82; dams on, 93; defined, 60, 81; flow controls, 89, 160; importance of, 61, 89–90; miles of, 145; number of fish species in, 84; originating water sources, 82; pollution sources, 89; Texas, 6, 82, 84–85
roseate spoonbill, 42, 123
rule of capture, 162
run, 82, 83
runoffs, 17, 89, 158
rushes, 106

Sabine River, 22, 82
Sahara, 15
Salado Springs, 76
salicornia, 106
salinity, 63, 116
saltwater, 5, 35
saltwater fish: adaptations, 29, 43, 121; dependence on wetlands, 62, 104; fishing for, 140, 147, 148–50; number in Texas, 32
San Antonio, 76–77
San Antonio Bay, 116, 117, 123
San Antonio River, 24
San Antonio Springs, 76
sand crabs, 132
sand dunes, 132
sand sheet wetlands, 105
San Felipe Springs, 77
San Marcos River, 86
San Marcos Springs, 19, 74, 76, 86
San Pedro Springs, 76–77
San Solomon Springs, 70
saturated zone, 18
scales, 31
scavengers, 50, 51
science journals, Texas Aquatic Science curriculum, 170
scrapers, 98
seabirds, 135
seafood, economic importance of, 64, 121, 140, 141. *See also specific types*
seafood production, estuaries, 125
seagrass, 35, 119–20, 122
seagrass beds, 63–64, 119–20, 123, 125, 140
seagrass habitat, 121

INDEX

197

seasonal weather patterns, 13
sea turtles: adaptations, 135; environmental threats, 142; green sea, 138; growth in numbers of, 141; Gulf of Mexico, 135, 138–40; habitats, 132, 136; Kemp's ridley, 43, 125, 138–40, 166; loggerhead, 138; protections for nesting, 138–39
seaweed, 35, 142
second-order streams, 85–88
sedges, 106
sediment, 24–25, 99, 111, 131
sedimentation, 24–25, 83, 98
seeps, 104, 133, 141
senses, fish, 32
sensory adaptations, 73
sewage treatment facilities, 20
sharks, 135, 136
shellfish. *See specific types of*
shiners, 74
shipping lanes, 140
shoalgrass, 120
shoreline erosion, coastal wetlands, 63, 64
shredders, 87–88, 98
shrimp: economic importance of, 64, 140, 141; as fish bait, 147; in the food chain, 123; protective habitats, 35, 118–19
sight in fish, 32
silt, 8
slot limits, 151
sloughs, 104
smell, fishes sense of, 32, 39
snails, 85, 88
snow, 17
snowflakes, creation of, 12–13
snowmelt, 17
South Texas wetlands, 105
spawn, 123
species: competition for resources, 53; defined, 28; invasive, 55–56, 75, 136; niche, 47; nonindigenous, 55; pollution-sensitive/-tolerant, 88–89

sperm whale, 138
sport fisherman, 145
spotted bass, 88
spotted seatrout, 119, 121, 123
spring ecosystems, 59, 60, 71, 73–75
spring-fed, wetlands, 105
springs: defined, 60, 71; economic importance of historically, 76–77; formation of, 71; life from underground aquifers in, 74; number in Texas, 71; number without flow, 165; Texas, 73, 77. *See also specific springs by name*
stakeholders, 164
stocked, 100
stonefly, 88, 89
storm surges, 117
stormwater runoffs, 17
stream banks, 82–83
stream channel, 82, 83, 85
stream ecologist, 91
stream ecosystems, 59, 60–61, 84–85
stream fish, 84–85
stream flow, 81, 89
stream plants, 83–84, 85
streams: allocating water to, 163–64; aquatic community, 86–88; biodiversity in, 84–85; channelizing, 111; dams on, 93; defined, 60, 81; ephemeral/intermittent, 23, 82; first- and second-order, 85–88; fish species in, 84, 85; higher-order, 88; importance of, 61, 89; influences on size and flow, 82; originating sources, 82; parts of, 82–84; perennial, 23; riparian rights, 163; sources of pollution in, 89; temperature variation in, 84; Texas, 22, 81, 82, 84–85; third- through fifth-order, 88; tributary, 22; water quality indicators, 88–89
striped bass, 97
structural trait, 28
submergent plants, 97
subsidence, 162

suckers, 88
sunfish, 88, 147
surface water, 16–17, 19, 159, 161–63
surface water allocation, 161–62, 163
survival of the fittest, 54–55
swamps, 104
swim bladders, 30, 38, 85
swimming pools, 75, 76

tail fins, 30, 31
TEKS (Texas Essential Knowledge and Skills), Texas Aquatic Science curriculum alignment with, xi, 173
temperature: effect on pollution, 8; fish spawns and, 48; range, in Texas, 15; variation in streams, 84; water's role in, 15
temperature zones, lakes, 95–96
Texas: climate, 14; population growth, 164; rainfall in, 7, 14, 82; state freshwater fish, 39; water available, 6–7; water law in, 159, 161–63; weather in, 13–16. *See also specific bodies of water*
Texas Aquatic Science, vii, 171–72
Texas Aquatic Science curriculum: about the text, 169; assessment, 171; field investigations, 171; importance of outdoor learning, 172; lessons, 170; science journals, 170; teacher guide overview, 169–70; teaching resources website, vii; TEKS alignment, xi, 173
Texas Aquatic Science Project, xi–xii
Texas blind salamander, 38, 74, 75
Texas Commission on Environmental Quality, 10, 158, 160
Texas Department of State Health Services (TDSHS), 121, 155
Texas Essential Knowledge and Skills (TEKS), Texas Aquatic Science curriculum alignment, xi, 173
Texas Forest Service, 160
Texas General Land Office, 160
Texas Marine Mammal Stranding Network, 166
Texas Natural Resource/Environmental Literacy Plan, 172
Texas Parks and Wildlife Department (TPWD): Angler Recognition Program, 152–53; career opportunities, 10; fishing regulations, 151; Guadalupe bass restoration efforts, 39; responsibilities of the, xi–xii; wetlands protection and restoration, 110
Texas State Soil and Water Conservation Board, 158, 160
Texas Stream Team, 153, 158
Texas Water Development Board, 160
Texas wild rice, 74, 86
third-order streams, 88
threatened species, 39
tidal wetlands, 106
tides, 117–18, 135
toothless blindcat, 74
tourism, 100, 112, 125, 135–37
toxic algae blooms, 121, 135
Trans-Pecos wetlands, 105
transpiration, 12
tributary streams, 22
Trinity River Authority, 160
trophic levels, 51–52
tuna, 135
tunicates, 138
turbid, 8
turbulent water flow, 82
turtlegrass, 120
turtles, 40, 107, 135

unconfined aquifers, 70
unicellular, 71
United States, amount of water available in the, 6
Uvalde, 77
Uvalde County Underground Water Conservation District, 160

vertebrates, 30
volunteer beach cleanups, 166

wastewater treatment facilities, 19–20, 82, 89
wastewater treatment plant workers, 20
water: demand for, 5, 164; forms of, 2; hot, 8; human need for, 5; importance of, 1–2; planning for future needs, 164–65; properties of, 2–4, 8; regulatory agencies, 160. *See also specific types of, e.g.* surface water
water cycle. *See* hydrologic cycle
water flow, turbulent, 82
waterfowl, 42, 109–10, 114, 123–25
water law: Clean Water Act, 10, 26, 158; legal rights to own and use, 159, 161–63
water lily, 106, 108
water molecules (H_2O), 2–3
water pennies, 88
water pump, solar-powered, 12
water quality: defined, 9; improving, 111; indicators of good, 88–89; monitoring agencies, 160; in nature, 4, 12; testing, 9, 10; watershed, 24. *See also* pollution (water)
water quality regulator, 10
water right permits, 161–62, 163
watershed, 21–22, 24
watershed action planning, 159
watershed address, 22–23
watershed management, 99
water table, 18, 79, 111
water temperature: effect on pollution, 8; fish spawns and, 48; variation in streams, 84
water vapor, 12

waves, 132, 134
weather, 13–16, 150
well water, 19
West Galveston Bay, 120
West Indian manatee, 138
wetlands: biodiversity, 103, 106–109; Central Texas, 105; ciénegas, 105; coastal, 63, 64, 103, 104, 106, 109–10, 116, 124–25; defined, 62, 102; destruction of, 112, 114, 125, 165; East Texas, 105; forested, 105; formation of, 67–68; freshwater, 104; Gulf Coast, 106; healthy, 103; High Plains, 105; importance of, 110–11, 112, 114; oxbow lakes creating, 105, 106; playa lakes, 70, 105; pollution filtering, 111; resacas, 105; riparian, 105; sand sheet, 105; smell of, 103; South Texas, 105; spring-fed, 105; tidal, 106; Trans-Pecos, 105
wetlands conservation, 110, 114
wetlands ecosystems, 59, 62, 103, 106–109
wetlands habitat, 118–19, 123
wetlands tourism, 112
whales, 138
whale shark, 135, 136
whirligig beetle, 107
whooping crane, 51, 53, 123, 125
widgeongrass, 120
wildlife technician and biologist, 114
winds: bays and estuaries, 126; fishing and, 150

zebra mussel, 56
zooplankton, 34–36, 54, 88, 96–97, 121, 131